LANDSCAPING
Principles and Practices
5th Edition

The Residential Design Workbook

Online Services

Delmar Online
To access a wide variety of Delmar products and services on the World Wide Web, point your browser to:
    http://www.delmar.com/delmar.html
    or email: info@delmar.com

thomson.com
To access International Thomson Publishing's home site for information on more than 34 publishers and 20,000 products, point your browser to:
    http://www.thomson.com
    or email: findit@kiosk.thomson.com

A service of I(T)P®

LANDSCAPING
Principles and Practices
5th Edition

The Residential Design Workbook

Ferrell Bridwell

Africa • Australia • Canada • Denmark • Japan • Mexico • New Zealand • Philippines
Puerto Rico • Singapore • Spain • United Kingdom • United States

NOTICE TO THE READER

Publisher does not warrant or guarantee any of the products described herein or perform any independent analysis in connection with any of the product information contained herein. Publisher does not assume, and expressly disclaims, any obligation to obtain and include information other than that provided to it by the manufacturer.

The reader is expressly warned to consider and adopt all safety precautions that might be indicated by the activities herein and to avoid all potential hazards. By following the instructions contained herein, the reader willingly assumes all risks in connection with such instructions.

The Publisher makes no representation or warranties of any kind, including but not limited to, the warranties of fitness for particular purpose or merchantability, nor are any such representations implied with respect to the material set forth herein, and the publisher takes no responsibility with respect to such material. The publisher shall not be liable for any special, consequential, or exemplary damages resulting, in whole or part, from the readers' use of, or reliance upon, this material.

Cover Art: Courtesy of Jack E. Ingels
Cover Design: McKinley Griffen Design
Text Design: Stillwater Studio, Stillwater, New York

Delmar Staff:
Publisher: Tim O'Leary
Acquisitions Editor: Cathy L. Esperti
Senior Project Editor: Andrea Edwards Myers
Production Manager: Wendy A. Troeger
Marketing Manager: Maura Theriault

COPYRIGHT © 1997 Delmar, a division of Thomson Learning, Inc. The Thomson Learning™ is a trademark used herein under license.

Printed in the United States of America
6 7 8 9 10 XXX 03 02 01

For more information, contact Delmar, 3 Columbia Circle, PO Box 15015, Albany, NY 12212-0515; or find us on the World Wide Web at http://www.delmar.com

International Division List

Asia
Thomson Learning
60 Albert Street, #15-01
Albert Complex
Singapore 189969
Tel: 65 336 6411
Fax: 65 336 7411

Australia/New Zealand:
Nelson/Thomson Learning
102 Dodds Street
South Melbourne, Victoria 3205
Australia
Tel: 61 39 685 4111
Fax: 61 39 685 4199

Latin America:
Thomson Learning
Seneca, 53
Colonia Polanco
11560 Mexico D.F. Mexico
Tel: 525-281-2906
Fax: 525-281-2656

Spain:
Thomson Learning
Calle Magallanes, 25
28015-MADRID
ESPANA
Tel: 34 91 446 33 50
Fax: 34 91 445 62 18

Japan:
Thomson Learning
Palaceside Building 5F
1-1-1 Hitotsubashi, Chiyoda-ku
Tokyo 100 0003 Japan
Tel: 813 5218 6544
Fax: 813 5218 6551

UK/Europe/Middle East
Thomson Learning
Berkshire House
168-173 High Holborn
London
WC1V 7AA United Kingdom
Tel: 44 171 497 1422
Fax: 44 171 497 1426

Canada:
Nelson/Thomson Learning
1120 Birchmount Road
Scarborough, Ontario
Canada M1K 5G4
Tel: 416-752-9100
Fax: 416-752-8102

ALL RIGHTS RESERVED. No part of this work covered by the copyright hereon may be reproduced or used in any form or by any means—graphic, electronic, or mechanical, including photocopying, recording, taping, Web distribution or information storage and retrieval systems—without the written permission of the publisher.

For permission to use material from this text or product contact us by Tel (800) 730-2214; Fax (800) 730-2215; www.thomsonrights.com

Library of Congress Cataloging-in-Publication Data:
Bridwell, Ferrell M.
    Residential landscape design/Ferrell Bridwell.
      p. cm.
    ISBN: 0-8273-6539-X
1. Landscape design. I. Title.
SB472.47.B75   1996
712'.6-dc20
                            96-13301
                              CIP

Contents

Preface vi

Exercise
   1   Identifying Materials and Equipment 1
   2   Using Equipment and Working with Scale 11
   3   Developing Lettering Skills 17
   4   Using Symbols in Drawing the Residence 25
   5   Using Symbols in Drawing Trees 31
   6   Using Symbols in Drawing Shrubs 37
   7   Using Symbols in Drawing Ground Covers, Vines, and Flower Beds 43
   8   Drawing Symbols for Nonplant Features 49
   9   Understanding Foundation Plantings 55
  10   Recognizing Faulty Foundation Design 61
  11   Designing a Foundation Planting 67
  12   Organizing Space in the Landscape 77
  13   Making Thumbnail Sketches 83
  14   Designing Walks and Drives 89
  15   Drafting Guest Parking and Turnarounds 93
  16   Designing Patios 103
  17   Designing Decks 109
  18   Understanding Balance 115
  19   Designing Curvilinear Gardens 119
  20   Designing Geometric Gardens 125
  21   Understanding Focalization 131
  22   Organizing/Beginning the Planting Plan 139
  23   Placing Man-Made Features on the Plan 145
  24   Developing the Design 149
  25   Selecting Plants 155
  26   Labeling Plants on a Plan 159
  27   Preparing a Finished Plant List 163
  28   Preparing a Title Block 167

Preface

In recent years, our population has become increasingly more urbanized. The overwhelming majority of high school students attend schools located in urban and suburban communities, and college students seeking degrees in agriscience-related areas are increasingly from nonrural communities. While the population is mostly nonfarm, so are the skilled and professional careers in modern agriscience. Careers in horticulture have become very popular in suburban America to satisfy the needs and demands of homeowners and businesses. Home gardening, both vegetable and ornamental, has become an important national pastime.

As a horticultural skill, landscape design has become essential in helping property owners to create and enhance the livable space of their properties. Commercial properties and larger residential estates are often designed by professionals with a degree in Landscape Architecture. However, most residential landscapes are planned and installed by skilled workers with high school, technical school, or college training in landscape horticulture.

High school programs in agricultural education have begun to include more instruction and/or courses in ornamental horticulture. Such instruction should include skills development in landscape design. Instructors sometimes skip or minimize this important area because they lack a systematic approach to such instruction.

On the postsecondary or college levels, beginning design classes should accommodate students with little or no prior experience in drafting or design.

The purpose of this material is to provide a step-by-step guide to basic skills development in landscape design. The materials can be adapted to large-group instruction or to the individualized format presented. In either case, the student should master an exercise before moving to another

EXERCISE 1
Identifying Materials and Equipment

Objective

To familiarize students with the tools of landscape drafting and the use(s) of each.

Skills

After studying this unit, you should be able to:

- identify landscape drafting tools, by name, from line drawings or the actual tools.
- give the tool needed for a particular task or function.

Materials Needed*

Drawing board (minimum size 17" x 22")
T-square
Triangles (45-45-90 and 30-60-90)
Adjustable triangle
2H drawing pencil
Lead holder
Lead pointer or pencil sharpener
Vinyl eraser
Eraser shield
Engineer's scale (triangular)
Architect's scale (triangular)
French curves
Compass
Protractor
Circle template ($\frac{1}{16}$" to 1" circles, minimum)
Drafting paper (plain or gridded)

Pictures may be substituted for items not available.

Introduction

The landscape designer, like the architect or engineer, must be familiar with the use of basic drafting tools. There are literally thousands of products on the market today. However, the following is a description of basic tools with which the beginning student of landscape design should become familiar.

Drawing Board

The drawing board gives a smooth surface for drafting on paper. It can be used on a table, desk, or other steady surface. Drawing boards come in various sizes, and are made of smooth-sanded laminated wood or plastics. Wooden boards or tables should be covered with a vinyl drawing board cover. Such covers are more durable than wood and less subject to damage (see Figure 1-1).

Figure 1-1 Drawing board and parallel bar

T-square

A T-square is used for drawing straight lines that are parallel to the edge of the drawing board. It may be used for either vertical or horizontal lines. A T-square is not required where boards or tables are equipped with a parallel bar or drafting machine (see Figures 1-1 and 1-2).

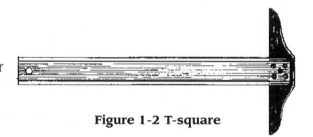

Figure 1-2 T-square

Triangles

Triangles are used to draw angled lines off of any straight line. The most commonly used angles are 45 degrees and 90 degrees. Angles other than 30, 45, 60, or 90 degrees can be drawn with the aid of a protractor or an adjustable triangle (see Figure 1-3).

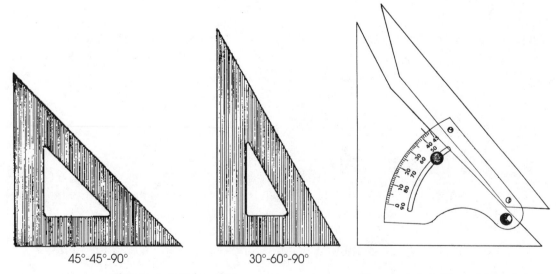

45°-45°-90° 30°-60°-90°

Figure 1-3 Triangles   **Figure 1-4 Adjustable triangle**

Adjustable Triangles

These triangles cost considerably more than standard fixed triangles, but can be used to measure or plot any angle between 0 and 90 degrees in one-degree increments (see Figure 1-4).

Protractor

A protractor is used to measure the angle of any two joining lines from 0 to 180 degrees, in one-degree increments for the standard protractor or 0 to 360 degrees for the round version. It is used to determine existing angles, or it can be used to mark nonstandard angles not available on triangles (see Figure 1-5).

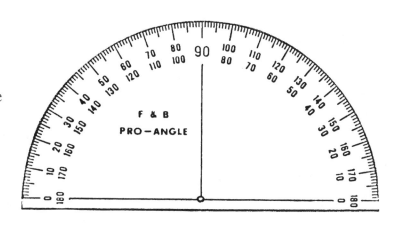

Figure 1-5 Protractor

Drawing Pencils

Drawing pencils come in varying degrees of hardness, usually 2B, B, HB, F, H, and 2H through 9H. Drafting is done with lead having an HB rating or higher. An H-rated pencil contains harder lead, and produces lighter lines which are less likely to smear. Avoid leads with a B rating, since B leads are much softer and are most often used for artwork or sketching. A good choice for beginning landscape designers is in the range HB to 2H (see Figure 1-6).

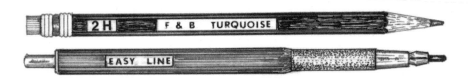

Figure 1-6 2H pencil and lead holder

Lead Holders

Lead holders, or mechanical pencils, are more expensive initially, but are much less expensive over the long run. The standard lead holder uses lead 2 mm thick and requires a lead pointer. Some prefer lead holders with smaller lead that are available in 0.3 mm, 0.5 mm, 0.7 mm, or 0.9 mm thicknesses. These holders do not usually require sharpening of the lead, but because of the smaller lead, breakage of the point is more frequent (see Figure 1-6).

Pencil Sharpeners and Lead Pointers

Drafting pencils may be sharpened with any quality standard sharpener, either mechanical or electric. Lead pointers are essential for lead holders using 2 mm leads. Most pointers come with replaceable carbide blades, and some allow for adjustment of point taper (see Figure 1-7).

Figure 1-7 Two popular styles of lead pointers

Erasers

Erasers should completely erase errors or changes without rubbing holes in the drafting paper. The best drafting erasers are made of vinyl and are nonabrasive. Electric erasers are faster, but are not essential for the beginning student of design (see Figure 1-8).

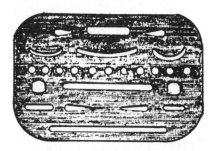

Figure 1-8 Vinyl eraser

Eraser Shields

Eraser shields are used to erase unwanted lines or parts of lines without smudging or erasing desired lines. Shields have various sizes and shapes of openings to allow one to match the line to be erased. Eraser shields are very inexpensive, and every draftsperson should have at least one (see Figure 1-9).

Figure 1-9 Eraser shield

Circle Templates

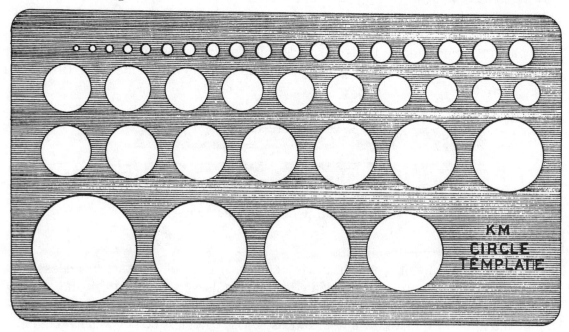

Figure 1-10 Circle template

Circle templates allow rapid drawing of circles to represent trees, shrubs, and other landscape features. Templates chosen should have a variety of circle sizes (see Figure 1-10). Templates appropriate for landscape should have circles from 1/16" to 1" minimum. A template with 1/16" to 2¼" circles is more useful. Extra large circles can be drawn with a compass.

Compass

A compass or bow compass is an adjustable instrument used to draw circles or arcs. It has two legs–one containing a metal point and the other containing a pencil or lead. The distance between the two legs is one-half the diameter of the circle. For example, a 2" setting will yield a 4" circle. Select a compass with a threaded adjustment bar to prevent slippage and maintain adjustment while drawing (see Figure 1-11).

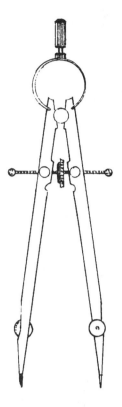

**Figure 1-11
Bow compass**

Scales

Scales are used for measuring or drawing lines that represent smaller dimensions on paper than in "real life." For example, a 1:10 scale means that 1" on paper will represent 10' of real space. Triangular scales are popular because each has six or more scales on one instrument. **Engineer's scales** contain scales of 1/10, 1/20, 1/30, 1/40, 1/50, and 1/60.

Architect's scales usually contain scales of ½, ¼, ⅛, 1/16, ⅜, ¾, and 3/16. Landscape designers commonly use scales of 1/10 or ⅛ (see Figure 1-12).

Figure 1-12 Engineer's scale (top) and architect's scale (bottom)

Lettering Guides

Lettering guides are used to draw guidelines essential to attractive lettering. The Ames Lettering Guide is an inexpensive, adjustable guide with many different spacings, both standard and metric (see Figure 1-13).

Nonadjustable guides usually have only four line widths, but they allow for rapid drafting of guidelines (see Figure 1-14).

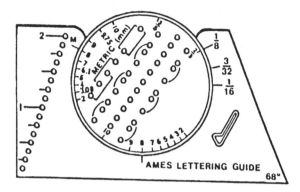

Figure 1-13 Ames Lettering Guide

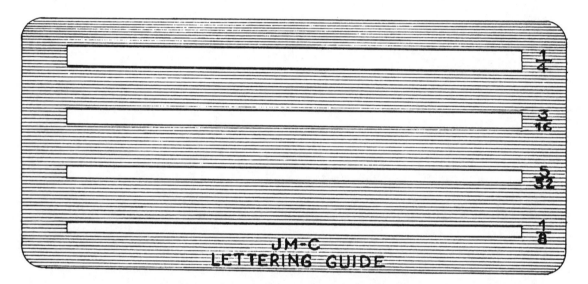

Figure 1-14 Nonadjustable lettering guide

French Curves

French curves serve as a guide in making professional looking curved lines for bed areas, walks, drives, etc. The degree of curve can be changed by moving the french curve to an edge or section of the sketch having the same degree curve. There are many curves available in many price ranges. A package of three or four is usually adequate for landscape design (see Figure 1-15).

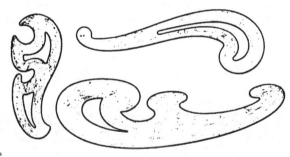

Figure 1-15 French curves

Drafting Machines

Drafting machines replace T-squares, parallel bars, and triangles, and allow for more rapid drafting. A drafting machine is not essential for beginning students, but students in advanced classes, draftspeople, or professionals will find it the single most time-saving mechanical device available (see Figure 1-16).

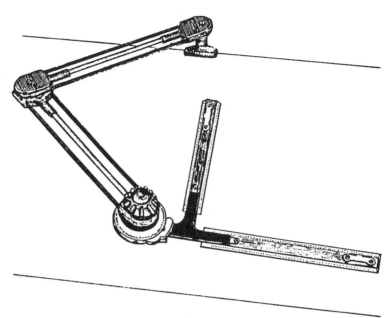

Figure 1-16 Drafting machine

Drafting Papers

Drafting papers are available as opaque or transparent in a wide variety of sizes. Opaque papers are fine for rough drafts or final copies where the drawings are not to be blueprinted. When blueprints are desirable, it is essential to draw on high-quality vellum made of 100% cotton and labeled as 100% rag. For ease in drafting, paper with nonreproductive blue grids is available as opaque or transparent. When blueprinted or photocopied, the blue lines do not reproduce. Gridded paper is usually available in 4 x 4 (¼" = 1'), 8 x 8 (⅛" = 1'), or 10 x 10 (¹⁄₁₀" = 1') (see Figure 1-17).

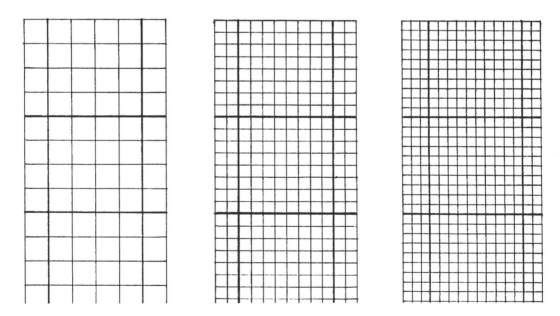

Figure 1-17 Gridded papers–4 x 4, 8 x 8, 10 x 10

Assignment

1. Study the reading material until you understand the use of each instrument.
2. Review the pictures included in this exercise or examine the actual instruments (if provided by your instructor).
3. Complete the evaluation exercise for Exercise 1 without the assistance of the written materials.

Evaluation

1. The teacher will score the evaluation exercise.
2. You should score 70 percent or higher before starting the next exercise.

Notes

Exercise 1 Evaluation

Student Name_____Date _____Score_____

Fill in the Blanks

Fill in the blanks with the best answers.

1. A 45-degree line can be drawn quickly with the aid of a _____.
2. _____ _____, _____ _____, and _____ are useful in drawing lines parallel to the edge of the drawing board.
3. A(n)_____ or a(n)_____is useful in drawing nonstandard angles such as 37 degrees.
4. A _____ _____ allows rapid drawing of circles to indicate trees and shrubs.
5. An H rating indicates that a pencil has _____ lead.
6. Do not use leads with a _____ rating when drafting.
7. _____ _____ give a smooth, professional look to curving lines.
8. The best drafting erasers are made of _____.
9. Very large circles may be drawn with a _____.
10. _____ _____ allow one to erase a line without erasing adjacent or adjoining lines.
11. An _____ scale contains scales that are multiples of 10.
12. A 1:20 scale means that one inch equals _____ _____.

Identification

For blanks 13 through 20, give the correct names for the corresponding items pictured on the next page.

13. Item A_____ 17. Item E_____
14. Item B_____ 18. Item F_____
15. Item C_____ 19. Item G_____
16. Item D_____ 20. Item H_____

10 Exercise 1

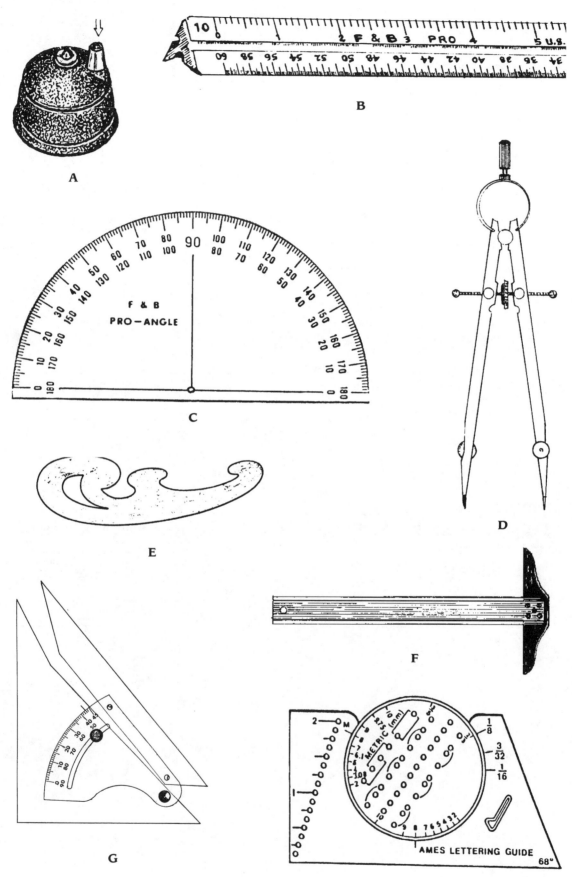

A

B

C

D

E

F

G

H

Exercise 2: Using Equipment and Working with Scale

Objective

To provide hands-on practice in the use of equipment and materials for landscape drafting.

Skills

After studying this unit, you should be able to:
- demonstrate a knowledge of scale by completing a basic scale drawing.
- convert a drawing on a 1:20 scale to 1:10, or convert a drawing on a 1:16 scale to 1:8.
- utilize various drafting instruments to complete the assignment.

Materials Needed

Drawing board
T-square
Triangles
Drawing pencil (HB, F, H, or 2H)
Vinyl eraser
Eraser shield
Engineer's scale or Architect's scale
French curves
Compass
Protractor
Drafting paper (with or without grids)
Drafting tape (not masking tape!)

Note: This exercise can be completed without a T-square and drawing board if paper with a fade-out grid is used.

Exercise 2

Introduction

A landscape plan must be accurate–just like the design of a building or any other design that utilizes space. In order to be accurate, it must be drawn to a chosen scale. Simply stated, scale is a miniature representation of the real thing. A landscape plan should indicate the exact number of plants needed, while giving a visual reference of the space needed in comparison to lawn areas or man-made features.

Before one can draw a landscape plan that will be accurate and usable, it is essential that one have experience with the correct use of appropriate tools. As you complete this exercise, it is essential that you be perfectly honest. If you feel you do not understand the proper use of a tool, ask your instructor for additional help or practice.

As you work with drafting tools, you will come to realize that scale is the basis of all such tools. One must understand scale first and foremost. All other tools exist to help you draft plans to scale with ease and accuracy.

The word "scale" is also used to describe the measuring instrument used in drawing. On the design level, one should refrain from using the word "ruler" in favor of the more descriptive word–"scale." There are two such scales used in landscape design–the engineer's scale and the architect's scale. Such scales are usually triangular, allowing for six different scales on the engineer's instrument, and eleven on the architect's.

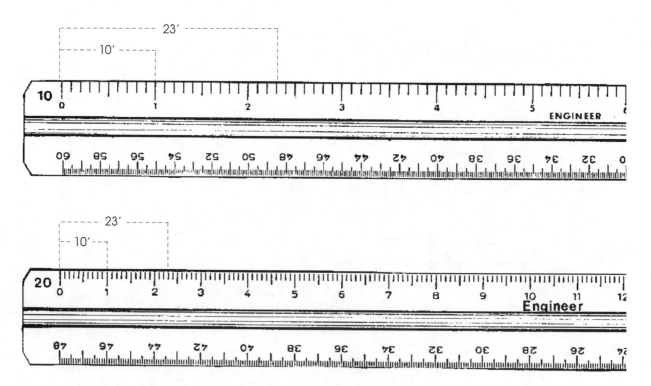

Figure 2-1 Engineer's scale

The engineer's scale is very popular with landscape designers. For most residential properties, a scale of 1:10 will allow the property and building(s) to be placed on a standard sheet of vellum drafing paper (17" x 22", 24" x 36", 30" x 42", or 36" x 48"). On larger residential properties, a scale of 1:20 may be used; however, some symbols become very small at this scale. A better choice would be to divide the property into two or more areas and use a 1:10 scale to draft the plan on two or more sheets of vellum.

A 1:10 scale means that 1" is equal to 10'. Another way of looking at this is to say that ⅒" is equal to 1'. In either case, every mark on the scale represents one foot. On the 1:20 scale 1" is equal to 20'. A drawing on this scale is exactly one-half the size of the same drawing on a 1:10 scale (see Figure 2-1).

You will note that for every 10', a longer mark exists on the scale, and a number is present. Just add 0 (zero) to the number on the scale. For example, on the 1:20 scale, 1 means 10' and 2 represents 20'.

Some designers prefer the architect's scale. The most popular architect's scale for landscape designers is the 1:8 where 1" = 8' or ⅛" = 1'. For larger properties, the 1:16 scale can be used, and on tiny properties a 1:4 scale is often appropriate.

As you study the architect's scale, you will note that the ⅛" scale and the ¼" scale are located on the same edge. The ⅛ inch scale reads left to right, while the ¼" scale reads right to left. This makes the ⅛" scale somewhat more confusing at first, but with practice you will learn to use one while ignoring the other (see Figure 2-2).

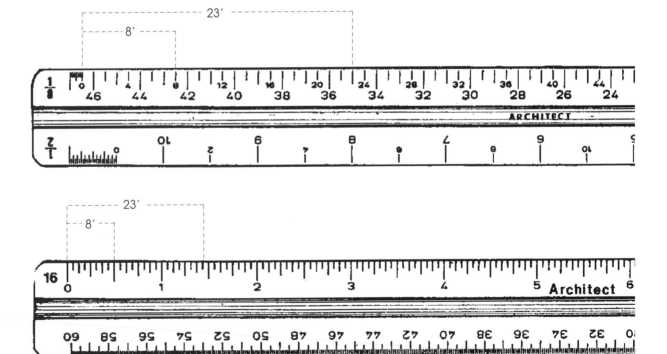

Figure 2-2 Architect's scale

Study the assignment drawing, examine your scale, and be sure you understand scale before starting the assignment.

Assignment

This assignment will enable you to practice scale and use some of the tools you studied in Exercise 1. The assignment is as follows:

1. Secure an 8½" x 11" sheet of drafting paper (standard vellum is fine if you have a drawing board and T-square). Tape all four corners to your drawing surface with the 11" side horizontal. If you are using 10 x 10 or 8 x 8 grid and a T-square, you must align the grid lines with the T-square.

2. The drawings included with this exercise are drawn on 1:20 and 1:16 scales. Draw the same residence on a 1:10 or a 1:8 scale.
   Note: Measure each line with the scale indicated, and then draw each line using the related scale as shown on the Exercise 2 assignment sheet.

3. Once your drawing has been evaluated as acceptable, *save it. It will be used in Exercise 3.*

Notes

Exercise 2 Evaluation

Student Name_____Date_____Score_____

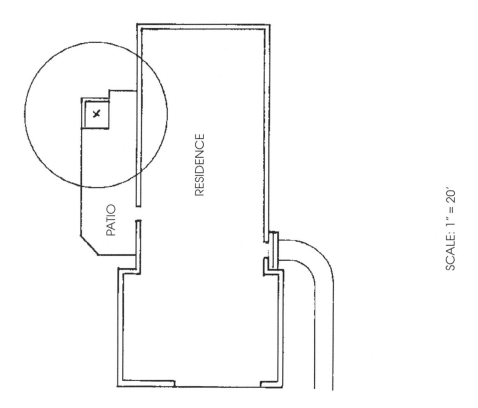

Evaluation

Consideration	Points	Student Score	Instructor Score
Exact scale achieved	75		
Lines smooth and consistent	15		
Drawing is neat	10		
Total	100		

15

Exercise 2 Evaluation

Student Name_____Date_____Score_____

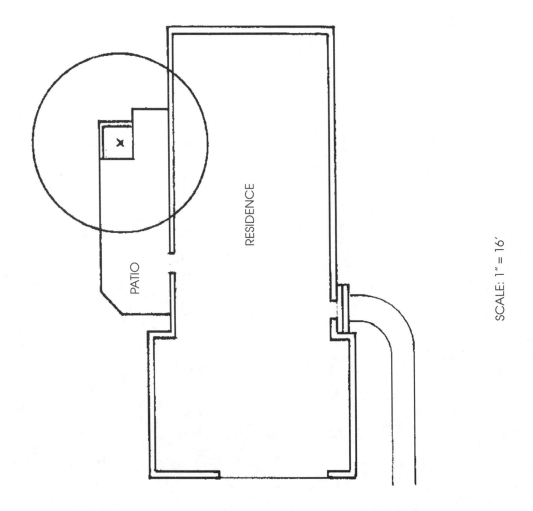

Evaluation

Consideration	Points	Student Score	Instructor Score
Exact scale achieved	75		
Lines smooth and consistent	15		
Drawing is neat	10		
Total	100		

16

Exercise 3: Developing Lettering Skills

Objective

To have students develop good lettering skills through supervised practice.

Skills

After studying this unit, you should be able to:
- properly use guidelines when lettering.
- draw letters of consistent height and spacing.
- draw letters using a single stroke.
- demonstrate your lettering style.

Materials Needed

Drawing board and T-square
Engineer's scale or architect's scale
Eraser and shield
Drawing pencil (HB, F, H, or 2H)
One 8½" x 11" sheet drawing paper
Drafting tape
House plan drawn in Exercise 2

Introduction

Good lettering is essential to give a professional or distinctive look to a drawing. It can "sell" both you and your plan to a prospective client. On the other hand, poor lettering is very noticeable on a drawing, and it can cause others to doubt your ability as a designer. Time spent developing your lettering skills is time well-spent.

Lettering is a developed skill. Although some people master lettering more quickly than others, everyone can develop acceptable lettering skills through practice. The following explanations should prove helpful.

1. Always use guidelines. Even the most experienced professionals use them. Horizontal guidelines are drawn with the aid of a T-square or, preferably, a lettering guide. To letter, you must have two guidelines with the spacing between them determining the size of the letters. First, draw the top guideline. Using a scale, measure down the desired distance and place a dot. Then, using the dot as a reference, draw the second guide. A lettering guide eliminates this procedure. Always draw guidelines lighter than your lettering, and never erase them after lettering. A blueprint or photocopy may pick up your guidelines, but that is acceptable so long as the lettering is darker.

    Vertical guidelines are used for margins, columns, or as a guide for centering. Refer to Exercises 27 and 28 for usage of vertical guidelines.

2. All lettering should be parallel with the bottom of the sheet. This enables one to read everything on the sheet without having to turn it.

3. Letters should touch both the top and bottom guidelines. This gives even height to the letters.

4. Letters are usually drafted in capitals; however, lowercase, even cursive letters, have been used effectively by architects and designers. Caution should be exercised in attempting to get too "fancy."

5. Don't go over a line twice. This makes that part of the letter darker, and it will stand out on the blueprint or photocopy. If you need to correct the letter, erase it fully and do it over.

6. Avoid wavy lines. Wavy lines are the result of marking too slowly. Relax and make deliberate strokes.

7. Avoid making letters too narrow or too wide. This will take practice, but will soon become second-nature (see Figure 3-1).

    Note: Use guidelines in your practice even if you are using fade-out vellum. Guidelines should become a habit with you, because there will be many times when you will have to letter on nongrid paper (see Figure 3-2).

LETTERS SHOULD TOUCH BOTH GUIDELINES

AVOID UneVen HEIGHT

TOO NARROW TOO WIDE

DON'T GO OVER LETTERS TWICE

USE SINGLE STROKES

Figure 3-1 Dos and don'ts of lettering

ABCDEFG
HIJKLMNO
PQRSTUV
WXYZ&123
4567890

abcdefghijkl
mnopqrstuvw
xyz

Figure 3-2 Sample lettering

Assignment

Tape an 8½" x 11" sheet of vellum to your drawing surface and align the bottom or top edge using your T-square. Complete the following assignment.

1. Draw several pairs of horizontal guidelines for several different size letters. Draw a vertical guideline for your left margin.
2. Draw two alphabets (including numbers) using 2/10" letters for engineer's scales or 3/16" for architect's scales. Draw the standard alphabet first, and then draw one of the slanting alphabets (see Figure 3-3).
3. Draw two alphabets for either 1/10" letters or 1/8" letters using the style you prefer.
4. Write your own name several times using both lettering sizes.
5. Write the sentence, "Good lettering skills are acquired through regular practice," several times (see Figure 3-3).
6. Using the house plan you drew for Exercise 2, label the rooms and areas as shown on the sample house plan lettering sheet (see Figure 3-4).
7. This assignment should be done over, if necessary, until it is acceptable.
8. Save your house plan. It will be used for Exercise 4.

Notes

ABCDEFGHIJKLMNOPQRST
UVWXYZ 1234567890 &
abcdefghijklmnopqrstuvw
xyz

ABCDEFGHIJKLMNOPQRSTUVWX
YZ   1234567890   &

ABCDEFGHIJKLMNOPQRSTUVWXYZ   12345
67890   abcdefghijklmnopqrstuvwxyz

ABCDEFGHIJKLMNOPQRST
UVWXYZ 1234567890 &
abcdefghijklmnopqrstuvwxyz

ABCDEFGHIJKLMNOPQRSTUVW
XYZ   1234567890   &

GOOD LETTERING SKILLS ARE ACQUIRED THROUGH REGULAR PRACTICE.

Figure 3-3 Various lettering sizes

Developing Lettering Skills 23

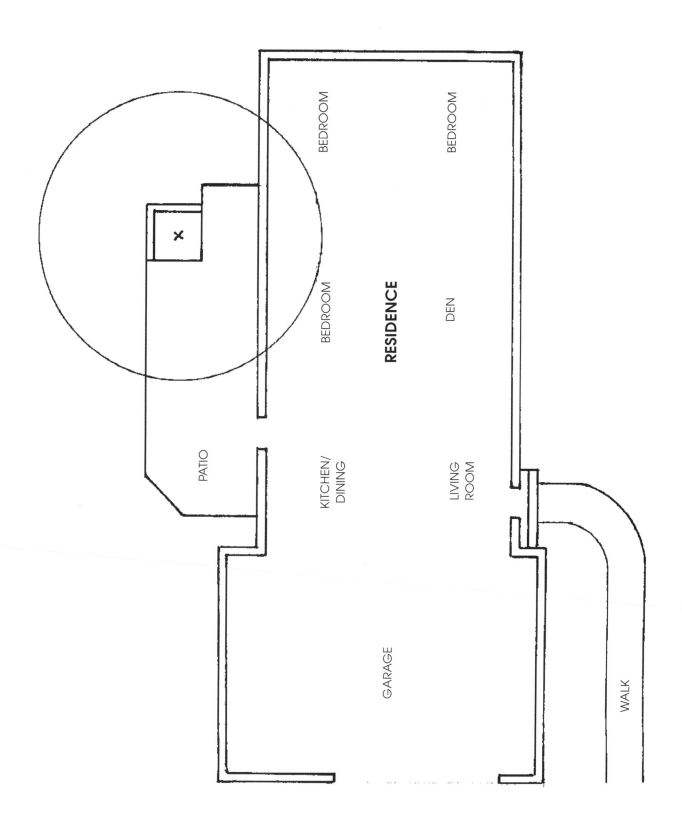

Figure 3-4 House plan lettering

Exercise 3 Evaluation

Student Name_____Date_____Score_____

Evaluation

Consideration	Points	Student Score	Instructor Score
Letters touch guidelines	50		
Letters are equal width	10		
Spacing is acceptable	10		
Pencil strokes are even	10		
Overall neatness	20		
TOTAL	100		

EXERCISE 4: Using Symbols in Drawing the Residence

Objective

To provide experience in drawing and recognizing symbols for ground covers, vines, and annual flowers on the landscape plan.

Skills

After studying this unit, you should be able to:

- draft walls, windows, and doors of a residence.
- draft surfacing materials for patios, walks, and drives of a residence.
- darken walls of a residence plan with uniform shading.

Materials Needed

House plan assignment sheet from Exercise 3
Drawing board and T-square
Engineer's scale or architect's scale
Eraser and shield
Drawing pencil (HB, F, H, or 2H)

Introduction

In drafting a residence, it is essential that the outside walls, windows, and doors be located accurately. This will determine the kind, number, size, and placement of the foundation plants (plants near the wall or foundation of a structure). To be usable, a landscape plan must be an accurate representation of the actual house and property.

It is not necessary to draw inside walls when drafting the residence; however, it is useful to indicate the location of rooms on the drawing as you did in Exercise 3. This is helpful to the designer in determining the views from a particular room or in locating other features in the landscape. For example, a satisfying or interesting view from the den or family room might be a primary consideration in developing the total garden design.

Once the general dimensions of a residence are drawn, the windows and doors should be the next consideration. If you are working from an existing architect's drawing, you can simply determine the size and location of features and draw them on your plan.

Start by measuring from a corner to the first window or door of the wall. Always check your measurement from both corners of a wall, and recheck often as you draw. Use the same procedure if you are taking measurements of the actual residence using a tape measure or other device.

The exact sizes of windows or doors vary greatly. For instance, a window might be 28 inches, 32 inches, or some other size. Since your engineer's scale does not show inches, you will have to estimate small measurements such as inches (see Figure 4-1A). A good rule-of-thumb for the landscape designer is to round off such measurements to the nearest one-half foot or 6 inches. It is best to round off to the next largest one-half foot. This will help ensure that plants which are proposed near windows will not cover or obstruct windows. The same procedure of rounding off to the nearest one-half foot can be used in drafting doors.

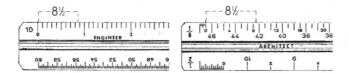

Figure 4-1A Estimating measurements

Standard windows can usually be drawn at 2½ to 3 feet. Bathroom windows are often smaller and measure 2 to 2½ feet. Picture windows and bay windows can be almost any size, but they usually vary from 6 to 8 feet in length. Doors usually vary between 2½ to 3½ feet.

Once windows and doors are located, the next step is to darken the remaining wall space. This serves to make the residence "stand out" on the plan, and it make windows and doors stand out. Strive to use the same degree or density (darkness) in shading the walls (see Figure 4-1B).

Figure 4-1B Locating window and doors and darkening wall space

Patios, drives, walks, and paths are made of concrete, stone, or brick. (Wooden decks will be covered in another exercise.) It is not necessary to draw symbols for the entire area. Drawing symbols to show the material for a section of the area is sufficient. Other surfacing materials, such as asphalt, can simply be labeled as such on the plan (see Figure 4-2).

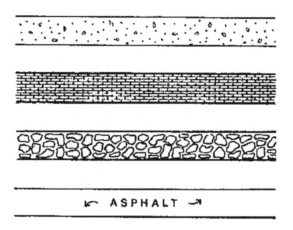

Figure 4-2 Various materials used for patios, drives, walks, and paths (continued on next page)

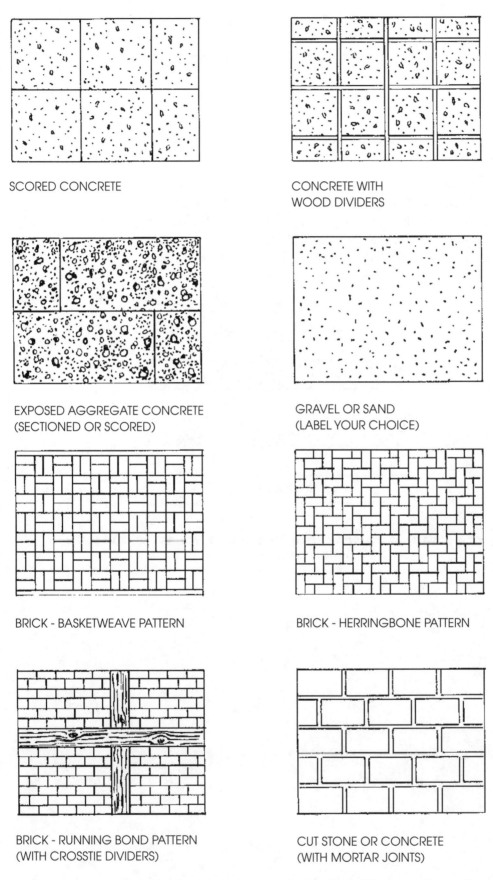

Figure 4-2 Various materials used for patios, drives, walks, and paths (continued from previous page)

Assignment

1. Study the Exercise 4 sample sheet to familiarize yourself with the symbols used.
2. Using the house plan drawn in Exercise 2 and lettered in Exercise 3, draw the windows in the same location as shown on the Exercise 4 Assignment Sheet.
3. Erase part of the wall connected to the patio to allow for the sliding glass door, then draft it as shown.
4. Darken the wall spaces between windows, as shown.
5. Using appropriate symbols, indicate the material you desire for the front walk and rear patio. You may select the concrete and flagstone drawn on the Exercise 4 Assignment Sheet, or you may select and draw another material.
6. Use a 1:10 scale or 1:8 scale.

Notes

Exercise 4 Evaluation

Student Name_____ Date_____ Score_____

Evaluation

Consideration	Points	Student Score	Instructor Score
Walls darkened and accurate	10		
Windows located accurately	30		
Doors located accurately	30		
Surfacing symbols accurate and acceptable	10		
Overall neatness	20		
Total	100		

EXERCISE 5

Using Symbols in Drawing Trees

Objective

To provide experience in drawing and recognizing symbols of trees on a landscape plan.

Skills

After studying this unit, you should be able to:

- draw symbols to represent small, medium, and large trees.
- draw symbols for both evergreen and deciduous trees.
- create symbols of your own.

Materials Needed

Drawing board and T-square
Drawing pencil (HB, F, H, or 2H)
Eraser and shield
Engineer's scale or architect's scale
Circle template (1/16" to 2¼")
One 8½" x 11" sheet of drafting paper
Drafting tape

Introduction

In designing landscapes, most permanent plants are represented by circular symbols. On smaller or partial plans containing a relatively small number of plants, it is appropriate to represent plants by circles drawn with a circle template. However, on larger plans with a greater number of plants, individual plants become lost in an endless "jungle" of simple circles.

The purpose of symbols is to represent variety in plants and, if desirable, to indicate whether the plant is deciduous (loses its leaves in fall) or evergreen (has leaves all year). Whereas perfect circles are not necessary in symbolizing plants, it is important to keep them mostly round.

Large trees are those trees that mature at 40 feet or taller. Medium trees mature at 20 to 40 feet, and small trees mature at 10 to 20 feet.

The following guidelines will work for the majority of trees in your plan:

1. Large trees   20 feet in diameter or greater
2. Medium trees   15 to 20 feet in diameter
3. Small trees   10 to 15 feet in diameter

To draw a tree symbol, decide the diameter circle you need. Choose a circle that is the desired size from the circle template. It might be helpful to place your scale over the circle template and measure various circles. If you do not have a circle large enough on your template, a compass can be used to draw the circle.

Once a circle is selected, draw it very lightly as you did in drawing guidelines for lettering. Next, decide upon a symbol and begin to draw the plant. You will note from the samples provided with this exercise, that evergreen symbols tend to have points or straight lines as part of the symbol. Deciduous symbols are less rigid or less pointed. Always mark the center of the plant, as this indicates the exact location of the initial planting (see Figures 5-1 and 5-2).

Assignment

1. Study the sample symbols provided. Notice differences between deciduous and evergreen symbols.
2. Obtain an 8½" x 11" sheet of drafting paper and tape it to your drawing surface.
3. Draw two deciduous and one evergreen for each size group for a total of nine trees.
4. Design your own symbols and draw one each of large, medium, and small trees for a total of three additional trees.
5. Be neat and strive for smooth, even lines. Do not hurry through this assignment. It only takes a little more time to make the drawings neat.
6. Use a 1:10 scale or a 1:8 scale.

Notes

34 Exercise 5

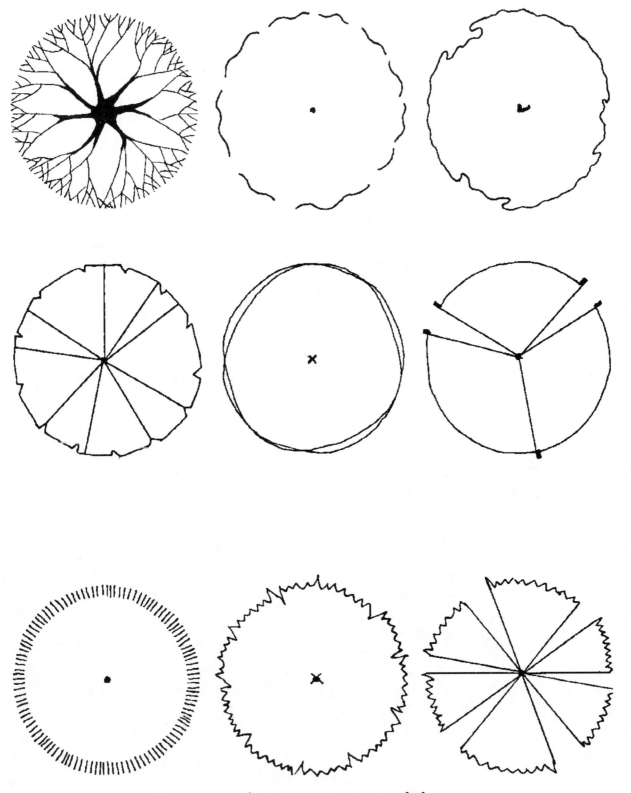

Figure 5-1 Large tree symbols

Using Symbols in Drawing Trees 35

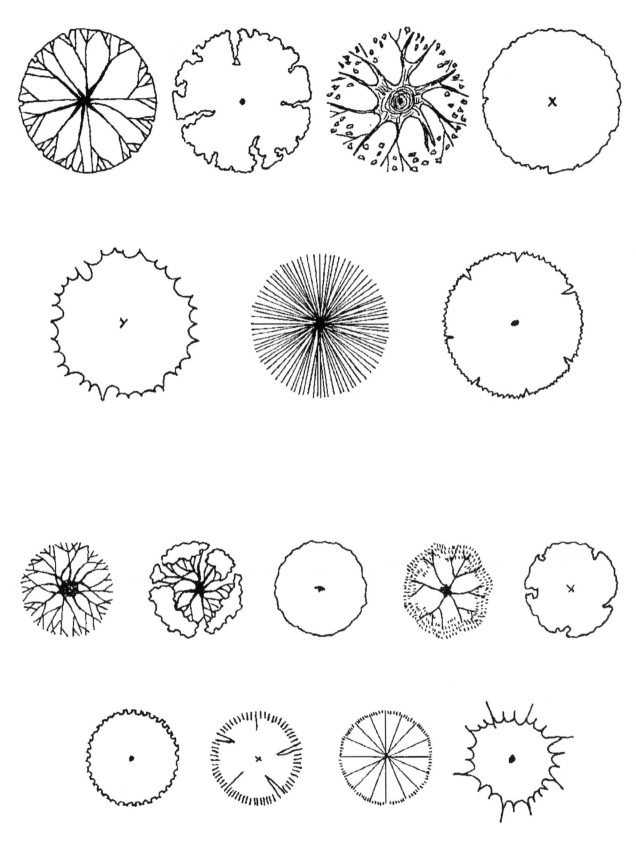

Figure 5-2 Medium tree symbols and small tree symbols

Exercise 5 Evaluation

Student Name_____Date_____Score_____

Evaluation

Consideration	Points	Student Score	Instructor Score
Measurements are accurate	50		
Symbols are uniform and circular	20		
Dots centered	15		
Overall neatness	15		
Total	100		

Using Symbols in Drawing Shrubs

Objective

To provide experience in drawing and recognizing symbols for shrubs on a landscape plan.

Skills

After studying this unit, you should be able to:

- draw symbols to represent large, medium, and dwarf shrubs.
- draw groupings of shrubs.
- create symbols of your own.

Materials Needed

Drawing board and T-square
Drawing pencil (HB, F, H, or 2H)
Eraser and shield
Engineer's scale or architect's scale
Circle template (¹⁄₁₆" to 2¼")
One 8½" x 11" sheet of drafting paper
Drafting tape

Introduction

With the exception of lawn grasses, shrubs usually make up the largest total number of plants in a landscape. Most shrubs are present in groups of the same species of plants or as a single plant that is part of a group. Plants that are part of a group must be drawn very accurately to prevent overcrowding in the actual landscape. This occurs when the symbols are drawn too small. On the other hand, if the symbols are drawn too large, the planting will appear too thin in real life. Therefore, one should draft the symbol to the size of a particular plant at mature width. This will require a source of information on landscape plants. There are many books that serve as a source of information on landscape plants. In addition, many free or inexpensive materials are available from the United States Department of Agriculture (USDA) or your local extension service. Reputable nurseries are usually happy to share information.

Shrubs vary in both width and height, but height is usually the measurement used in placing plants in size groups. **Dwarf shrubs** are plants that grow less than 4' in height at maturity. **Medium shrubs** grow 4 to 6 feet, and **large shrubs** are over 6 feet in height. Slight variations occur in different publications.

In the absence of specific information on width(s), the following guidelines will work for most plans:

1. Dwarf shrubs   3 to 4 feet in diameter
2. Medium shrubs   5 to 6 feet in diameter
3. Large shrubs   6 to 9 feet in diameter

When drafting the shrubs in a group, allow the symbols to touch or appear joined, as illustrated on the sample sheet. This will allow shrubs in the actual landscape to look "full," without being overcrowded (see Figure 6-1).

Using Symbols in Drawing Shrubs 39

Figure 6-1 Shrub symbols

Assignment

1. Study the sample symbols provided. Using your scale, measure some of the symbols drawn to get a feel for the sizes.
2. Tape a sheet of 8½" x 11" drafting paper to your drawing surface. Then draw three individual shrubs of each size group–dwarf, medium, and large–for a total of nine symbols.
3. Design your own symbols, one each for dwarf, medium, and large, for a total of three shrubs.
4. Draw a grouping of five shrubs for each size group.
5. Use a 1:10 scale or 1:8 scale.

Exercise 6 Evaluation

Student Name_____Date_____Score_____

Evaluation

Consideration	Points	Student Score	Instructor Score
Measurements are accurate	50		
Groupings are uniform	15		
Symbols neat in appearance	20		
Dots centered	15		
Total	100		

EXERCISE 7
Using Symbols in Drawing Ground Covers, Vines, and Flower Beds

Objective

To provide experience in drawing and recognizing symbols for ground covers, vines, and annual flowers on the landscape plan.

Skills

After studying this unit, you should be able to:

- draw symbols to represent both broadleaf ground covers and narrowleaf ground covers.
- draw symbols to represent annual flowers in the landscape.
- draw symbols to represent vines.
- understand the major purposes(s) of ground covers, vines, and annual flowers in the landscape.

Materials Needed

Drawing board and T-square
Drawing pencil (HB, F, H, or 2H)
Eraser and shield
Engineer's scale or architect's scale
French curve
45-45-90-degree triangle
Drafting tape

Introduction

Ground covers are permanent, low-growing plants that take the place of turf (grass) in a landscape. Almost every garden has an area that is suitable for some type of ground cover. Ground cover is easier to maintain than turf, and, with the modern popularity of ground cover, they are available in a wide variety of forms, leaves, flowers, color, and other characteristics. Ground covers are best used in combination with other landscape plants, especially trees. Steep embankments and shady areas under low-branched trees are ideal areas for ground covers.

In drawing ground covers, we do not show individual plants, since many are tiny compared to other plants. We show ground covers as a mass of plants, and we illustrate the shapes and boundaries of the bed areas. In using ground covers, a definite area should be established and maintained, since many of the plants spread by either roots or stems. Most can be contained by monthly maintenance shearing during the warm season. Figure 7-1 shows possible symbols for drawing ground covers.

Vines are used in the landscape in many ways. As with all plant groups, much variety exists. Vines are "climbing" plants which may or may not require support or attachment. Vines may be grown against a bare wall, on a fence, pole, or arbor (overhead beams) or trellis (upright latticework or crossed slatting). Carefully placed vines can add much interest and usefulness to the landscape.

Herbaceous flowers are used as accent plants. A bed planted in flowers will provide interest and color throughout warm weather. In designing areas for annual flowers in the landscape, it is not necessary to name every plant, since annual flowers may be changed from year-to-year or season-to-season. If perennial flowers are desired, more care in selection is required, since they emerge from the roots each spring. Flower beds should be planned in such a way that their absence will not detract from the landscape when they are not present. Symbols may or may not be necessary, but the areas should be labeled as *Flowers* (see Figure 7-1).

GROUND COVER

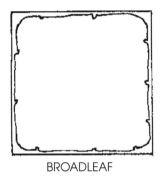
BROADLEAF

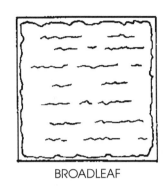
BROADLEAF

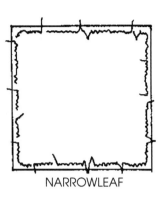
NARROWLEAF

VINES

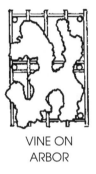
VINE ON ARBOR

VINE ON TRELLIS

VINE ON FENCE

VINE AGAINST WALL

FLOWERS

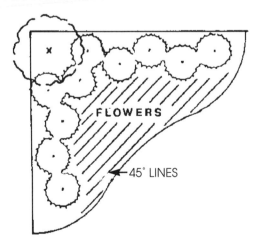

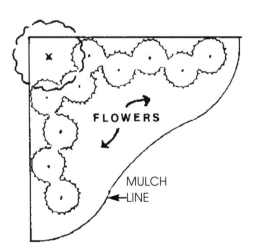

Figure 7-1 Symbols for ground cover, vines, and flowers

Assignment

1. Study Figure 7-1. Be sure you understand the symbols.
2. Follow directions, and complete the Exercise 7 Assignment Sheet. Draw directly on the sheet provided.
3. Use a 1:10 scale or a 1:8 scale.

Notes

Exercise 7 Assignment Sheet

Student Name_____Date _____

GROUND COVER

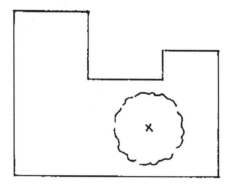 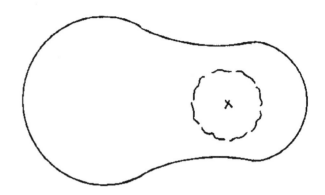

A: Use an appropriate symbol to show broadleaf ground cover in the bed area above.

B: Show narrowleaf ground cover in the entire space above.

VINES

A: Draw vines for the three sections of fence.

B: Draw a vine to cover the bare wall space above.

FLOWERS

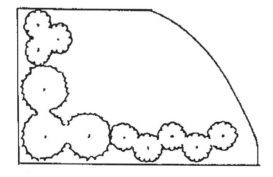 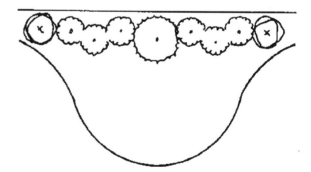

A: Use 45° lines to show a flower space in front of the shrubs. Label the space.

B: Label the bare space as flowers and draw arrows as shown on the sample sheet.

47

Exercise 7 Evaluation

Student Name_____ Date_____ Score_____

Evaluation

Consideration	Points	Student Score	Instructor Score
Symbols used are acceptable	30		
45 degree lines are accurate	30		
Lettering is acceptable	20		
Drawings are neat	20		
Total	100		

Drawing Symbols for Nonplant Features

EXERCISE 8

Objective

To provide practice in drawing and understanding nonplant features in the landscape.

Skills

After studying this unit, you should be able to:

- recognize symbols used for nonplant features such as fences, statuary, play equipment, etc.
- draw symbols used for nonplant features.
- label nonplant features.

Materials Needed

Drawing board and T-square
Drawing pencil (HB, F, H, or 2H)
Eraser and shield
Engineer's scale or architect's scale
French curves
One 8½" x 11" sheet of drafting paper
Drafting tape

Introduction

In previous assignments, you learned to draw and symbolize a residence, patios, walks, and plants. There are many other nonplant features and man-made features in landscapes. In drawing symbols for these man-made features, it is important to visualize how the feature(s) would appear from overhead. Imagine you were in a hot air ballon 300 feet overhead. How would the outline of the feature appear?

Man-made and natural items come in many shapes and sizes. Since many have similar shapes, it is important that most of these features be labeled directly on the drawing. The rule-of-thumb is that if you are in doubt as to whether others will understand what the symbol represents, then you should label it. Smaller size letters are recommended in labeling.

Always draw the features in exact scale. Just as a shrub must be drawn to mature size, nonplant features must be accurate in order to show the exact space that the feature will occupy in the actual landscape (see Figures 8-1 and 8-2).

Study Figures 8-1 and 8-2. Notice the drawing and lettering for each feature. If you don't understand a symbol, ask your instructor for additional explanation. The samples are drawn on a 1:8 scale, but you may use either the 1:10 or 1:8 scale.

Drawing Symbols for Nonplant Features 51

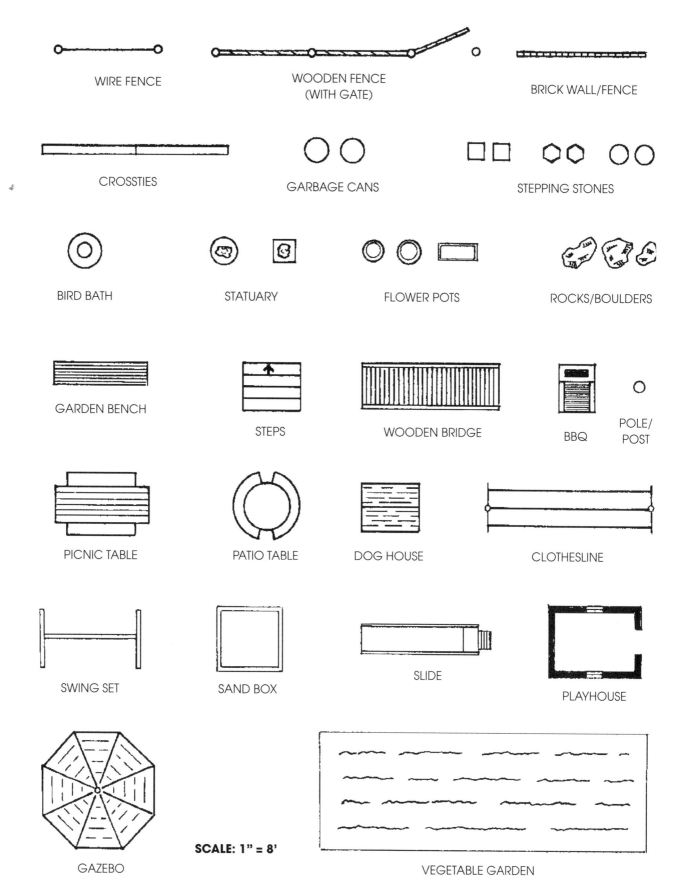

Figure 8-1 Nonplant symbols

Exercise 8 Evaluation

Student Name_____Date_____Score_____

Evaluation

Consideration	Points	Student Score	Instructor Score
Drawings are exact scale	40		
Symbols are appropriate	30		
Lettering neat and accurate	20		
Overall neatness	10		
Total	100		

Understanding Foundation Plantings

EXERCISE 9

Objective

To help students understand the importance and relationships of foundation plantings.

Skills

After studying this unit, you should be able to:

- understand the purpose and need for foundation plantings.
- recognize things to avoid in foundation plantings.

Materials Needed

Figures 9-1 and 9-2
Pen or pencil

Introduction

Foundation plantings are those plants used near or against the residence or other buildings. They are used to tie the residence to its outdoor environment, and such plants serve as the primary transition in this function. Foundation plants include those plants that are a few feet away from buildings but are a part of the foundation "bed" area. For example, you might have dwarf shrubs against the building, a bed of ground cover in front of the shrubs, and a small tree growing in the bed area of ground cover. The entire bed area should contain mulch with a distinctive edge at the point where the bed area and lawn area join.

Foundation plants are the first plants considered in a design. In order to understand foundation plantings, it is necessary to understand the shortcomings of earlier years. Earlier in this century, the public's knowledge of landscaping was greatly limited. "Bushes" were often chosen for some desirable feature, such as flowers, or some practical use, such as edible fruit, with little regard for mature size, deciduous or evergreen type, or relationships to other plants. As many people have become more affluent and have located in suburban communities, greater emphasis has been placed on designing landscapes that are attractive, functional, and low-maintenance.

Figure 9-1 presents some of the more common deficiencies in design. A very common shortcoming is the "toy soldier" effect. This landscape type utilizes one species of landscape plants, often round, which are spaced equally with noticeable gaps between plants. This type of design is monotonous, boring, and lacks creativity.

The **overgrown effect** involves the use of plants that are too large for the rooflines or windows of the residence. This dwarfs the home, and, in addition, requires much maintenance to control size.

The **crowded effect** is a large mass of confusion as a result of plants being placed too close at the time of planting. This is done to give the landscape an instant "fullness," and plants eventually lose their individual identity.

The **clipped effect** is when all plants are given a regular "haircut" and maintained with a very smooth edge. This results in many plants losing their unique growth habits.

The **unbalanced effect** results when too many plants, or larger plants, occur on one side or at the end of the planting. The landscape appears "tilted" and is obviously out of balance.

Finally, the **hedge effect** is when foundation plants are trimmed to a continuous box shape. This lacks in variety, and it gives the foundation no relief from the horizontal lines. Hedges are more appropriately used in borders as a living fence.

Understanding Foundation Plantings 57

"TOY SOLDIER" EFFECT

OVERGROWN EFFECT

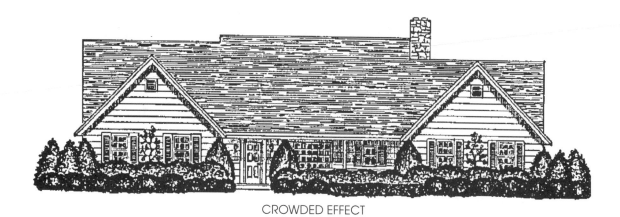
CROWDED EFFECT

Figure 9-1 Examples of poorly planned foundation plantings

CLIPPED EFFECT

UNBALANCED EFFECT

HEDGE EFFECT

Figure 9-1 *Continued*

Assignment

1. Study the reading material and samples until you understand the concepts.
2. Complete the Exercise 9 Evaluation without the assistance of the written materials or pictures.

Evaluation

1. Your instructor will score the written material.
2. Attain a score of 70 percent or higher before starting the next exercise.

Notes

Exercise 9 Evaluation

Student Name_____Date_____Score_____

Fill in the Blanks

Fill in the blanks with the best answers.

1. Foundation plants serve as a transition between the residence and the _____ _____.

2. Foundation plants may extend several feet from the residence if they are a part of the foundation _____ _____.

3. Modern landscapes should be designed so that they are _____, _____, and _____.

4. In the _____ effect, foundation plants are trimmed into a continuous box shape.

5. A landscape appears tilted in the _____ effect.

6. When plantings are _____, the plants lose their individual identity.

7. Selection of plants with no attention to mature size can result in the _____ effect.

8. The hedge effect gives no relief from the _____ lines of the residence.

9. The clipped effect can destroy the unique _____ habit of individual plants.

10. Using one species of plant, spaced evenly, with gaps between, is called the _____ _____ effect.

11. Plants are given a regular, smooth "haircut" in the _____ effect.

12. _____ _____ are the first plants considered in a landscape.

60

EXERCISE 10
Recognizing Faulty Foundation Design

Objective
To help students recognize errors in design and to propose solutions.

Skills
After studying this unit, you should be able to:
- understand solutions to design shortcomings, including those you studied in Exercise 9.
- give one or more names for each area.

Materials Needed
All sheets from Exercises 9 and 10
Pen or pencil

Introduction

The material included below will help you to understand how to overcome the "faults" that were studied in Exercise 9, as well as a few new concepts. In designing the foundation planting, one should strive for attractiveness, while ensuring that the landscape serves useful purposes and is easy to maintain.

1. Use taller plants on corners to "soften" the vertical lines while giving relief from the horizontal lines of the residence. The rule-of-thumb is that the corner plant(s) should mature at no more than ½ to ¾ the height of the corner. This is measured from ground level to the lowest point of the "boxing" or overhang. With a one-story house on a flat lot, this will average 8 to 10 feet. Houses with higher foundations will, of course, have taller corners. Two-story homes vary greatly, but usually can handle taller plants.

    Normally, medium-size shrubs of 4 to 6 feet in height can be used on corners of one-story residences, while large shrubs of 7 to 12 feet can be used on the corners of two-story residences. Following these guidelines will help avoid the overgrown effect.

2. Use lower-growing or dwarf plants under windows. For most windows there is a dwarf shrub of some species that will provide foliage without violating the overgrown effect. In some instances, very low windows can be planted with low-growing ground covers, since they vary from a maximum of 2 feet down to only a few inches. Another advantage to such treatment of windows is that it assures that there will be variety in the plants.

3. Show plants on the drawing at mature size. (For mature sizes, see Figure 6-1.) In drawing the plants, allow them to barely touch or almost touch. This will give a mass effect without being overcrowded. In addition, this will avoid the toy soldier effect illustrated in Figure 9-1.

4. Maintain balance. If the home or part of the home is **symmetrical,** each side of the symmetrical parts should be landscaped alike. (Symmetrical means that all features and measurements are identical on either side of the front door, which is centered.) If the home is unequal or **asymmetrical,** try to use equal amounts of foliage (leaves) on each end or side of the home. If the home is heavier on one end due to a shorter roof or recessed walls on the other end, use a larger amount of foliage on the light end in order to bring the overall effect into balance (see Figure 10-1).

5. If more than one row of plants is used in the foundation planting, place the taller-growing plant nearest the building and place the lowest-growing plants to the front. This will prevent plants from being hidden in the planting.

6. Always use evergreen plants near the wall. Deciduous plants can be harsh when foliage is absent. Planting them near the wall would enhance the harsh effect.

7. Use long, sweeping curves for borders between lawn and planting beds. Avoid choppy curves which are difficult to mow, and, in addition, look confusing.

8. Use three or more heights of plants in the foundation planting. This will give variety and a more interesting design while avoiding the hedge effect.

9. Use medium or taller growing dwarf plants on either side of the entry to focus attention to the entry.

10. Repeat some of the same plants on each half of the residence to give an organized look. However, repeating plants does not mean an exact repetition, except in symmetrical landscapes.

Some of these guidelines may require additional thought and research. Use any reference materials for additional study, and ask your instructor for additional help as needed.

Recognizing Faulty Foundation Design 63

SYMMETRICAL

PARTLY SYMMETRICAL – OVERALL ASYMMETRICAL

ASYMMETRICAL

Figure 10-1 Foundation design requires balance to be maintained

Assignment

1. Review all material from Exercise 9.
2. Study Figures 10-1 and 10-2. Study the Exercise 10 Assignment Sheet and try to identify as many faults as possible in Figure 10-2.
3. Complete the Exercise 10 Assignment Sheet. Assume that it is a one-story home with windows 3 feet off of ground level.

Evaluation

Turn in your completed assignment to the instructor for scoring. Achieve a score of 70 percent or higher before starting Exercise 11.

Notes

Recognizing Faulty Foundation Design 65

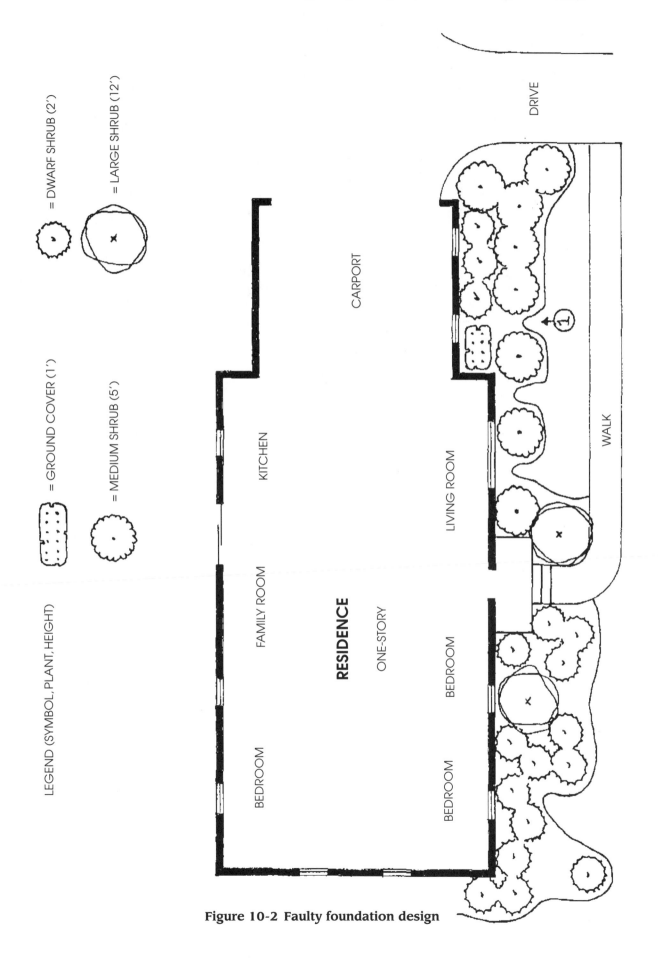

Figure 10-2 Faulty foundation design

Exercise 10 Assignment Sheet

Student Name_____ Date _____

In Figure 10-2, there are 8 or more faults, but you are responsible for finding 7.

To complete the assignment, place a number (2 through 8) near the fault, and draw an arrow to the fault. Beside the same number below, describe the fault. Under the solution column describe how the problem can be solved.

One fault has been labeled and explained for you as number 1. Find 7 additional faults and complete the assignment, using the example below as a guide.

Fault

1. Border between lawn and bed area is too choppy.

2. _____

3. _____

4. _____

5. _____

6. _____

7. _____

8. _____

Solution

Eliminate small curves. Use long, sweeping curves.

EXERCISE 11
Designing a Foundation Planting

Objective

To develop a foundation design.

Skills

After studying this unit, you should be able to:
- apply the principles you learned in Exercises 9 and 10 to develop an acceptable design.
- understand the relationship of "plan view" to elevation view.

Materials Needed

Drawing board and T-square
Drawing pencil (HB, F, H or 2H)
Engineer's scale or architect's scale
Eraser and shield
Circle template
French curves
Figures 9-1 and 10-1

Introduction

In completing this exercise, you will utilize the information presented here, as well as skills from previous exercises, to develop your own foundation design. In designing the foundation, it is important to visualize what the design will look like in real life. For beginning students of design, it is usually helpful to draw an **elevation view** that corresponds to the "plan view." The elevation view is simply a front view of the residence as seen from the front yard. This will involve more time, but the time will be well-spent as you begin to associate the plan to real life. The sample drawings in Exercise 9 are elevation views. As you gain more experience in design and learn more about

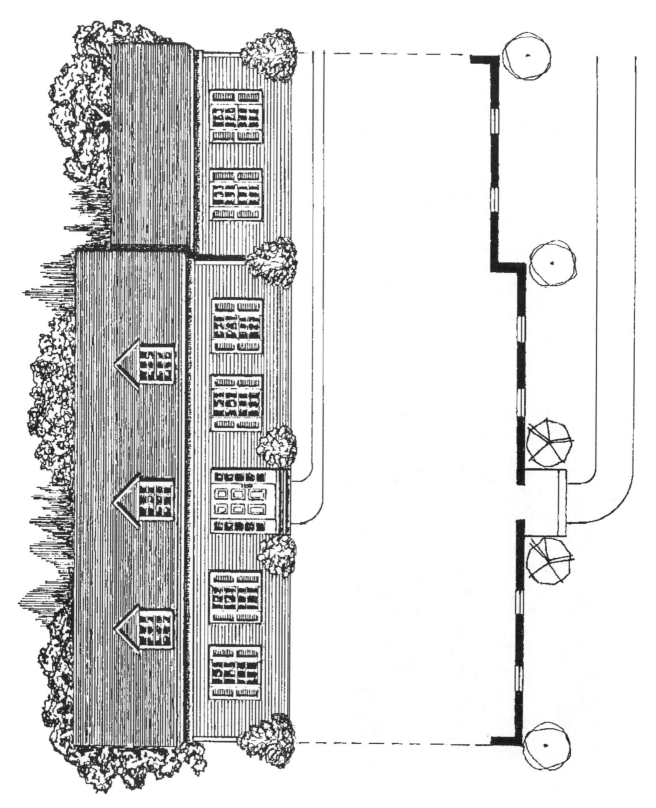

Figure 11-2 Foundation Planting Step 2

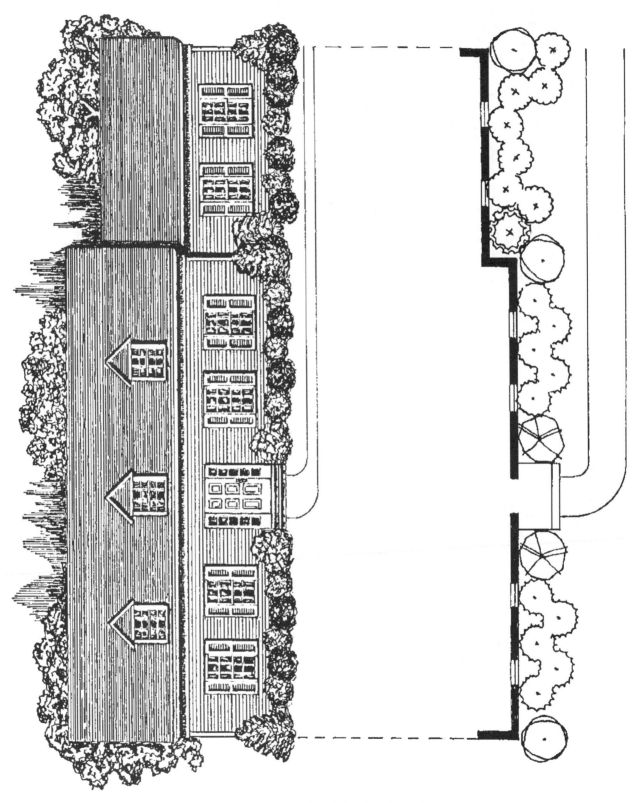

Figure 11-3 Foundation Planting Step 3

72 Exercise 11

Figure 11-4 Foundation Planting Step 4

Designing a Foundation Planting 73

The addition of plants, appropriately spaced and of appropriate mature size, will greatly improve the appearance and value of the property. However, as previously stated, one must consider the size of the structure and yard. Also, it is best to give some consideration to other gardens in your neighborhood (see Figure 11-4).

Side or rear foundations often continue the same technique. This might be important to an end or side that is highly visible. However, the rear landscape is strictly for family enjoyment and will vary greatly from family to family.

Some examples of ideas for the rear foundation, other than shrubs, might include:

1. Full-length patios, decks or terraces.
2. Herb gardens or cut flower beds.
3. Rose gardens or perennial flower plots.
4. Play areas for children.
5. Other family enjoyment features that do not block the view of the rear garden.

Assignment

1. Review Exercises 9 and 10 before starting your design.
2. Complete your design directly on the Exercise 11 Assignment Sheet.
3. Make both "plan view" and elevation drawings using one of the Assignment Sheets. Use Assignment Sheet A if you are using a 1:10 scale or Assignment Sheet B if you are using a 1:8 Scale.
4. Refer to sample sheets for Exercises 5 and 6 for sizes of plants.

Notes

Exercise 11 Assignment Sheet A

Student Name_____Date_____Score_____

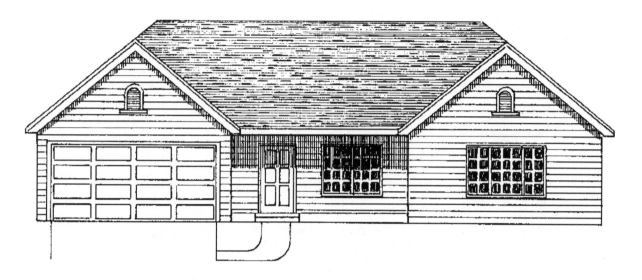

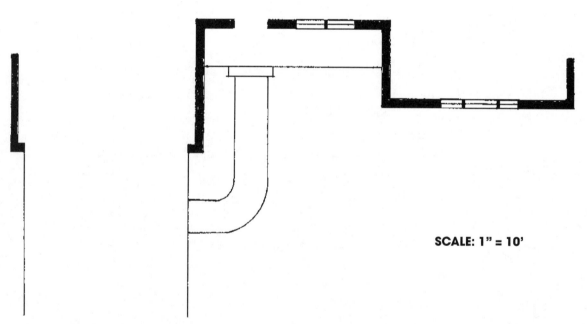

SCALE: 1" = 10'

Evaluation

Consideration	Points	Student Score	Instructor Score
Plants drawn to mature scale	20		
Medium shrubs on corners	20		
No large gaps exist	15		
All plants are correct height	15		
Design is balanced	25		
Total	100		

Exercise 11 Assignment Sheet B

Student Name_____ Date_____ Score_____

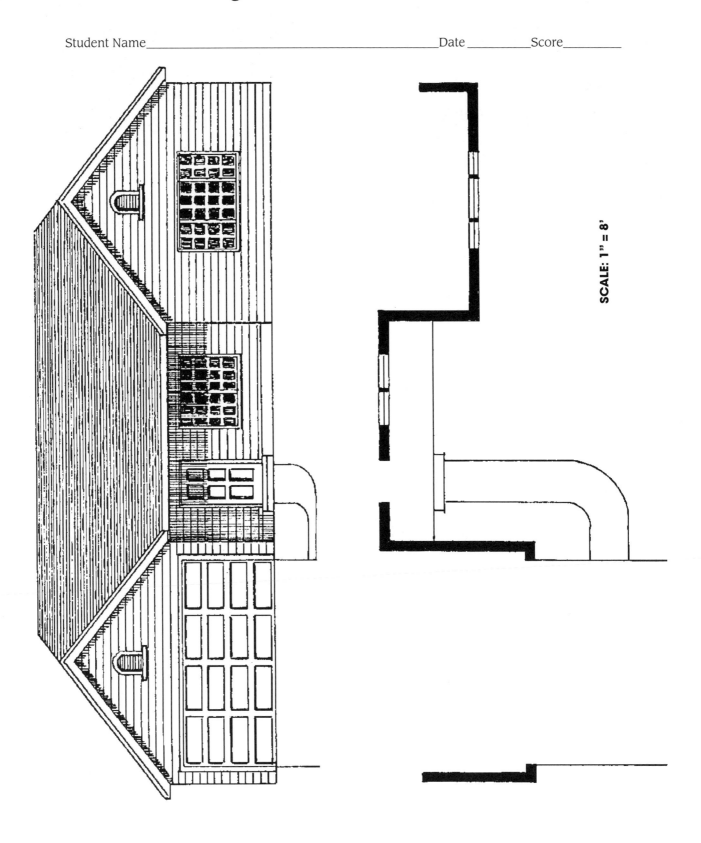

75

Exercise 11 Assignment Sheet B

Student Name_____Date _____Score_____

Evaluation

Consideration	Points	Student Score	Instructor Score
Plants drawn to mature scale	20		
Medium shrubs on corners	20		
No large gaps exist	15		
All plants are correct height	15		
Design is balanced	30		
Total	100		

EXERCISE 12
Organizing Space in the Landscape

Objective

To help students understand the reasons and benefits of well-planned organization of space in the residential landscape.

Skills

After studying this unit, you should be able to:
- understand the need for organizing space.
- give one or more names for each area.
- identify features for each area.

Materials Needed

Figure 12-1
Pen or pencil

Introduction

One of the more noticeable shortcomings of an "average" landscape is lack of planned organization of the space. This lack of organization can lead to unnecessary inconvenience, less functional space, and lack of attractiveness. As previously stated, a landscape should be both useful and attractive. Part of the challenge to landscape design is to maximize attractiveness after the usefulness has been planned. In order to understand usefulness and attractiveness, it is helpful to divide the landscape into functional units (areas). Refer to Figure 12-1 to help you define the following areas.

The **public area,** or entrance area, is the area(s) from which passersby will view the residence. This is almost always the front yard. Often, a side yard is part of this area, especially on some corner lots. If a family is limited on funds for landscaping, the public area is usually given the greatest priority. Almost everyone wants to give a favorable public impression.

78 Exercise 12

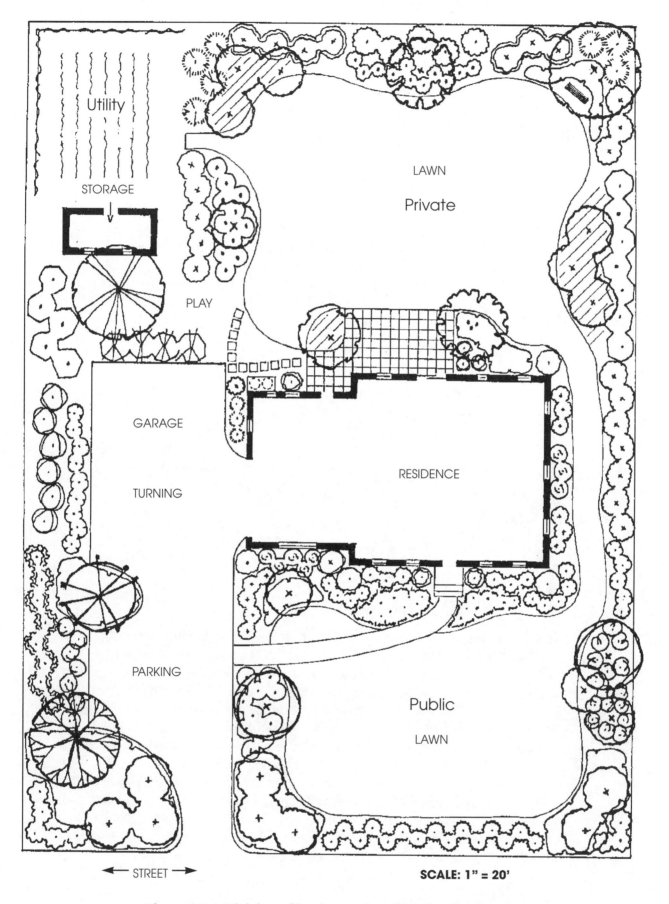

Figure 12-1 Division of landscape into functional units (areas)

The public areas should be simple, or, in other words, it should not be complicated. For example, the front walk should lead to the front door without taking visitors on a cross-country hike. Likewise, the guest parking should provide ease in maneuver while providing easy access to the front entry. The public area includes lawn, foundation plants, walks, and drives/parking. The public area should not include recreation equipment, play equipment for children, swimming pools, etc. Avoid the **carnival effect** of objects such as cheaper plastic animals, painted rocks, or tree trunks, or overuse of weird or unusual specimen plants.

The **private area,** sometimes called the living area, family area, or outdoor living room, is usually located in the backyard or rear garden. Sometimes all or part of a side yard is included. The private area need not be a barricade or completely isolated area, although it might. Depending on the nature or interests of the population, dwarf plants may serve simply to identify property boundaries. The purpose of this area is to provide an outside extension of the private living area inside the home–a king of warm weather retreat.

In addition to people's need for privacy, security is of utmost importance to today's urban population. Plants, fences, or both can be used to make the area private.

The private area should contain the space and arrangement necessary to meet the needs and desires of the family. In its simplest form, the private area will include a deck or patio, and area of open lawn, and plants that provide an attractive view. On a more complex level, the private area can include swimming pools and other athletic facilities, barbecue or picnic facilities, trails, "view" gardens, reflection pools, gazebos, and many other recreation or relaxation features.

The **utility area,** often referred to as the service area or work unit, is also located in the rear garden or side garden. In planning for utility, it is necessary to use the minimum amount of space to fulfill family needs and desires. Such economy of space will free more space for the private area. Some examples of features found in utility areas are vegetable gardens, utility buildings, workshops, pet facilities, greenhouses, clotheslines, compost bins, firewood, fuel tanks, and garbage containers.

The utility area should be screened from the private area, or, at least, the view should be softened by planting beds. Wherever possible, this area should be located on the driveway side of the yard for ease in access. Time spent designing the layout for the "service" items will prove highly rewarding, as well as visually increasing the size of the remaining spaces.

Finally, some families can benefit from a planned play area for children. Such an area can be planned so as not to detract from the beauty of the private area.

The **play area** should be located adjacent to, or as a part of, the private area. It is critical that the area be visible from the den or kitchen window, and that there is easy access to a rear entry door. Mulch, fine gravel, sand, or other materials provide more practical surfaces for play areas. Grass is almost always worn and unattractive under play equipment. A play area should contain a variety of playground equipment such as swings, slides, sandboxes, etc. A shade tree will result in better play area utilization by children during hot weather.

Assignment

1. Study and review Figure 12-1.
2. Complete the Exercise 12 Evaluation without the aid of the text or sample sheet.
3. Achieve a score of 70 percent or higher before starting the next exercise.

Evaluation

The instructor will score the Exercise 12 Evaluation.

Notes

Exercise 12 Evaluation

Student Name_____ Date _____ Score _____

Fill in the Blanks

Fill in the blanks with the best answers.

1. The three primary areas of a landscape are _____, _____, and _____.
2. A _____ area is needed for families with small children.
3. Dividing the yard into areas encourages the planned organization of _____.
4. A landscape should be both _____ and _____.
5. The front yard is almost always the _____ area.
6. The public area should be _____ rather than complicated.
7. The private area is sometimes called the _____ area.
8. The private area fulfills people's need for both _____ and _____.
9. _____, _____, and _____ are three features found in a private area.
10. The utility area is sometimes called the _____ area or _____ unit.
11. When planning for utility, one should use the _____ amount of space.
12. _____ and _____ are two features that could be found in a utility area of a residence.
13. A swimming pool should be located in the _____ area.
14. The utility area should be located on the _____ side of the yard.
15. A relaxation bench should be located in the _____ area.
16. A small vineyard should be located in the _____ area of the landscape.

81

EXERCISE 13

Making Thumbnail Sketches

Objective

To provide practical experience in sketching a property organizational plan.

Skills

After studying this unit, you should be able to:
- plan efficient use of space in a landscape.
- sketch informal diagrams of landscape areas.

Materials Needed

Exercise 12 sheets
Exercise 13 sheets
Drawing pencil (HB, F, H or 2H)
Eraser

Introduction

In planning the areas and space in a landscape, it is necessary to plan a specific location for each area and the individual features for each location. Since many changes take place in the development of a final plan, a "rough sketch" or **property organizational plan** is essential in getting started. This "rough sketch" does not identify individual plants or shapes for vegetable gardens, swimming pools, etc. It does identify the area most desirable for each component.

The "rough sketch" or property organizational plan is commonly called a **thumbnail sketch** because many of the somewhat circular sketches resemble a thumb-nail or a distorted bubble (see Figure 13-1). Another definition might be that of a **skeleton**, because it does provide the framework from which the completed design develops.

In developing the thumbnail sketch, it is important to use very light lines when working with the sheet of vellum that will be the final drawing. In such cases, thumbnails will be erased as the drawing progresses. One method often used is to place another sheet of tracing paper or vellum over your drawing and draw the thumbnails on it. This eliminates any damage to the original. Drawing on the original is far more convenient, provided you are very careful.

Study the house and property on the Exercise 13 Assignment Sheet. Begin visualizing your main area divisions. Before you begin your assignment, the following review will be helpful:

1. Draw thumbnail sketches for all surface areas of the property.
2. Include plant beds and lawn areas.
3. Label every thumbnail or bubble.
4. Draw lines that are lighter than normal. If you have trouble sketching with light pressure, a 6H pencil can be used.
5. Remember that exact scale is not necessary for this step. Also, it is not necessary to use guidelines for lettering, since the lettering will be erased before the final drawing is complete.

Making Thumbnail Sketches 85

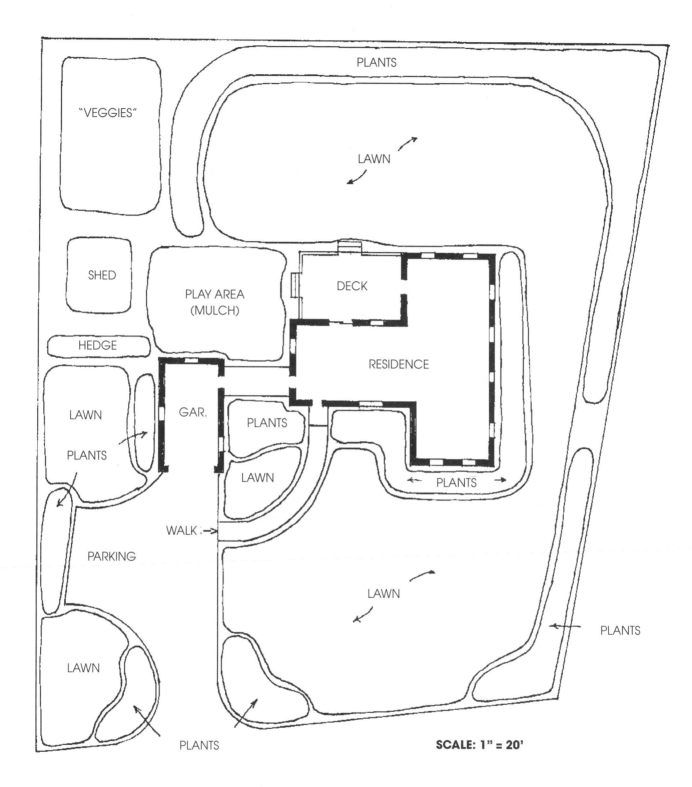

Figure 13-1 Thumbnail sketch

Assignment

1. Review Figure 13-1.
2. Complete a thumbnail sketch for the house and property directly on the Exercise 13 Assignment Sheet.
3. Assume that you will have the following features:
    a. garbage cans
    b. combination workshop/boat shed
    c. swimming pool
    d. vegetable garden
    e. play area for children
    f. firewood stack or pile
    g. neighbors on both sides
4. Notice that the scale is 1:20.

Notes

Exercise 13 Assignment Sheet

Student Name_____Date_____Score_____

PATIO

RESIDENCE

PARKING

SCALE: 1" = 20'

Exercise 13 Assignment Sheet

Student Name_____Date_____Score_____

Evaluation

Consideration	Points	Student Score	Instructor Score
All areas are "bubbled"	30		
Space is divided efficiently	30		
Overall plan is "workable"	20		
Lines drawn lightly	20		
Total	100		

EXERCISE 14

Designing Walks and Drives

Objective

To provide practice in drawing straight drives and front walks.

Skills

After studying this unit, you should be able to:

- draw a straight drive of appropriate size, length, and material.
- draw a front entry walk of appropriate size, length, shape, and material.

Materials Needed

Drawing board and T-square
Drawing pencil (HB, F, H, or 2H)
Engineer's scale or architect's scale
Eraser and shield
French curve
Sample sheets from Exercise 4

Introduction

Straight drives are very common for houses having carports or garages that open to the front yard. Under almost all conditions, such carports can accommodate a relatively straight drive from off the street. Very seldom is a curving drive necessary or desirable. On very large estates, some degree of curve might add interest, if not overdone. (Don't take visitors on a Sunday afternoon cruise of your front yard.)

Most garages are either single or double. A rule of thumb would be to allow a minimum of 10' for each car. For instance, a single car garage would have a drive of no less than 10' wide, and a double car garage would have a drive with a minimum of 20' width. Less frequently, you will find garages to accommodate three or more cars. Drives can be made of many different materials. Concrete and asphalt are more common. (Refer to the Sample Sheets for drawing or labeling surfacing materials.) Gravel is a temporary surfacing material and should be treated as such.

There are basically three types of walks that can be used in a residential design. The front walk or **entry walk** is designed to provide a comfortable walking surface for visitors and other guests. In addition, it should serve to direct such visitors to the front door of the residence. It may be straight or it may curve, depending on conditions. For instance, on very small properties where visitors park on the street, a straight walk might be the most practical. However, in most newer subdivisions, the walk runs from the drive to the front door. Some curving is usually necessary. Unnecessary curving will mean unnecessary walking, and it will tempt others to walk on the lawn. The front walk should be a minimum of 4 feet to allow two people to walk side-by-side. A variety of surfacing materials are available. Concrete is most common.

A **secondary walk** is a walk designed primarily for family members. It is placed near side or rear entries where family traffic is heavy. For example, a commonly used technique is to have such a walk from the drive to a rear entry door. This walk can be smaller; however, a minimum of 2 feet is desirable. Stepping stones are often used for secondary walks.

A **garden path** is sometimes useful in occasional movement throughout the garden. Such paths are usually more attractive in the overall design when they are not made of solid materials. A path made of mulch, lawn grass, or decorative pebbles would be appropriate. Size will vary depending on need, but 3 feet should be adequate.

Assignment

1. Draw (design) a drive for the residence on the Exercise 14 Assignment Sheet. Use Assignment Sheet A if using a 1:10 scale or Sheet B if using a 1:8 scale.

2. Draw a front entry walk that originates at the drive and runs to the front entry. Leave sufficient space for the foundation plants.

3. Draw symbols on a small section of both the drive and walk to indicate surfacing material.

Exercise 14 Assignment Sheet A

Student Name_____Date_____Score_____

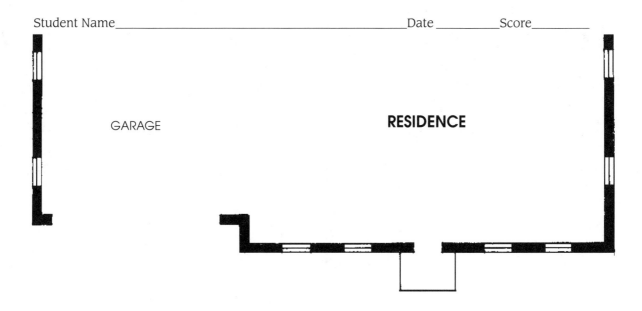

SCALE: 1" = 10'

Evaluation

Consideration	Points	Student Score	Instructor Score
Drive is correct size	20		
Walk is correct width	20		
Surface material is symbolized	20		
Drive and walk are appropriate	30		
Overall neatness	10		
Total	100		

91

Exercise 14 Assignment Sheet B

Student Name_____Date_____Score_____

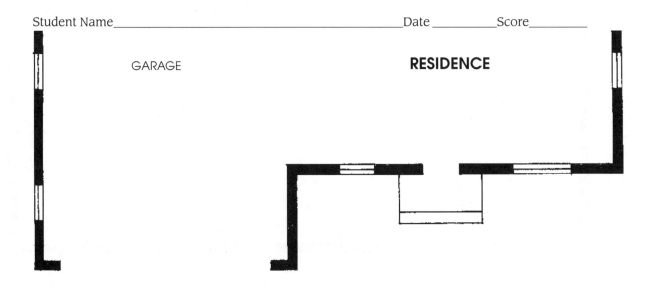

 STREET

Evaluation

Consideration	Points	Student Score	Instructor Score
Drive is correct size	20		
Walk is correct width	20		
Surface material is symbolized	20		
Drive and walk are appropriate	30		
Overall neatness	10		
Total	100		

EXERCISE 15
Drafting Guest Parking and Turnarounds

Objective

To provide practical experience in drafting guest parking and/or turnarounds in the landscape.

Skills

After studying this unit, you should be able to:

- understand the need for guest parking and turnarounds.
- identify different styles of guest parking areas and turnarounds.
- understand the measurements and angles used for parking and turning.
- select an appropriate parking and/or turnaround for a given residence and draft it to scale.

Materials Needed

Drawing board and T-square
Drawing pencil (HB, F, H, or 2H)
Eraser and shield
Engineer's scale or architect's scale
45-degree triangle or adjustable triangle
French curve(s)
Compass

Introduction

The residential family of today very often has two, or even more, passenger cars. The need for well-planned space to accommodate family parking has never been greater. Add to this two or three guest cars, and parking becomes even more complicated. Imagine that you need to run an errand and one or two cars have you blocked in. You must either move the cars yourself or have a family member or guest do so. Advance planning could have prevented this situation.

Older neighborhoods, especially those near the inner city, allow for street parking of guests. Such residences have a front walk that leads to the street. More modern suburban developments do not accommodate street parking, and the front walk leads to the drive.

The first consideration is family parking. This might be a garage or a surfaced area near one side that allows quick access to the residence. A multi-car family might have both. It is desirable to have a turnaround, where possible, to avoid having to back into a street.

Guest parking should accommodate a minimum of two cars. A socially active family might allow for more parking of guests. Guest parking should always be located in the front yard or one side of the front yard, with a front walk that leads guests to the front door.

Study the examples in Figures 15-1 through 15-4. The measurements given are minimum measurements, so it is acceptable to use slightly larger dimensions. Notice that turning areas give an "R" measurement. The "R" stands for radius, and it represents the smallest radius acceptable for turning an average family car. Notice the minimum turning radius for a 90-degree turn is 20'. To draw a curve of 20' R, set the steel point and pencil point a distance equivalent to 20' on the scale you are using.

Practice drawing a 20' R. Once you understand Figures 15-1 through 15-4, begin the Exercise 15 Assignment.

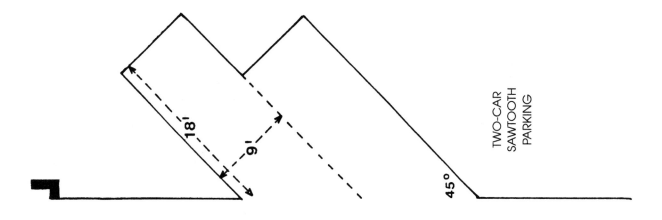

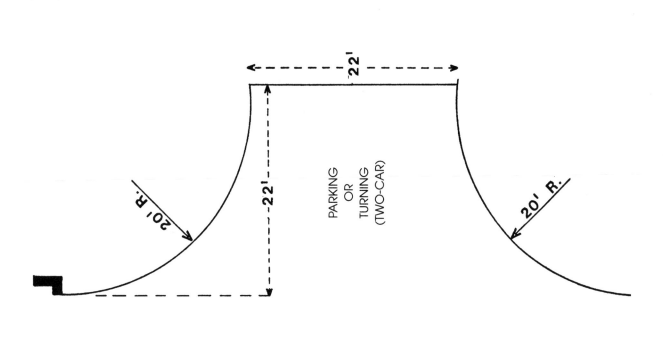

Figure 15-1 Garage entry from front

Exercise 15

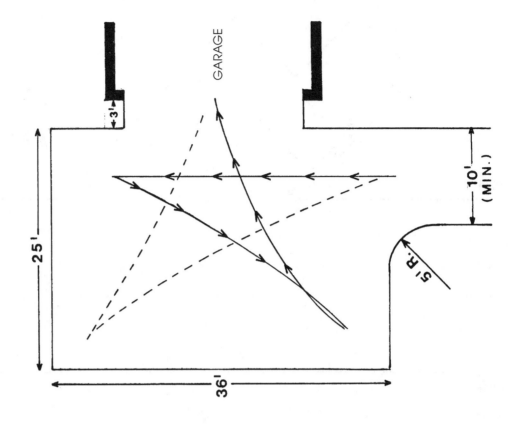

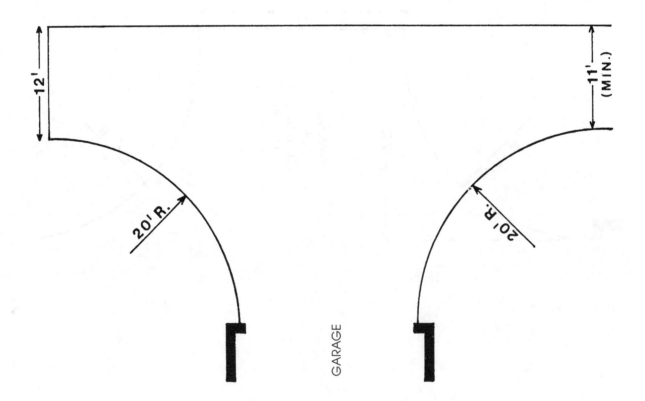

Figure 15-2 Garage entry from side

Drafting Guest Parking and Turnarounds **97**

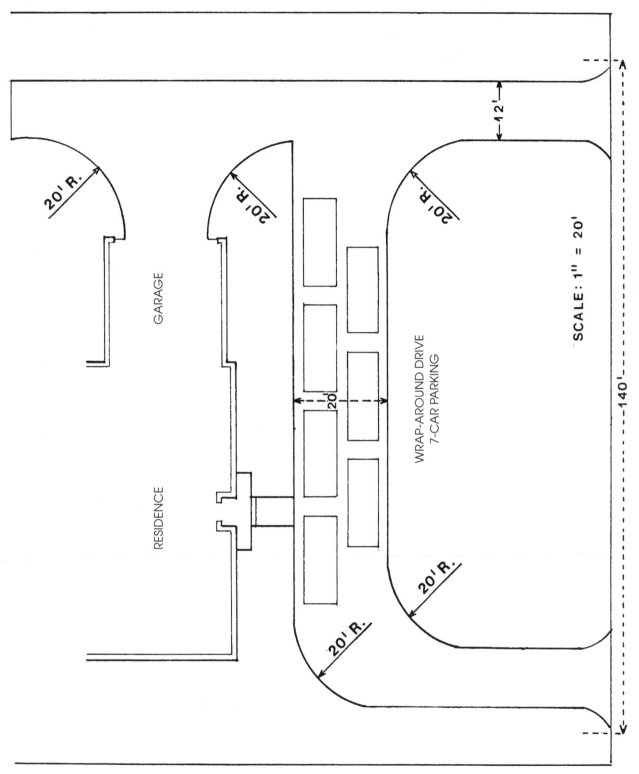

Figure 15-3 Large wrap-around

Exercise 15

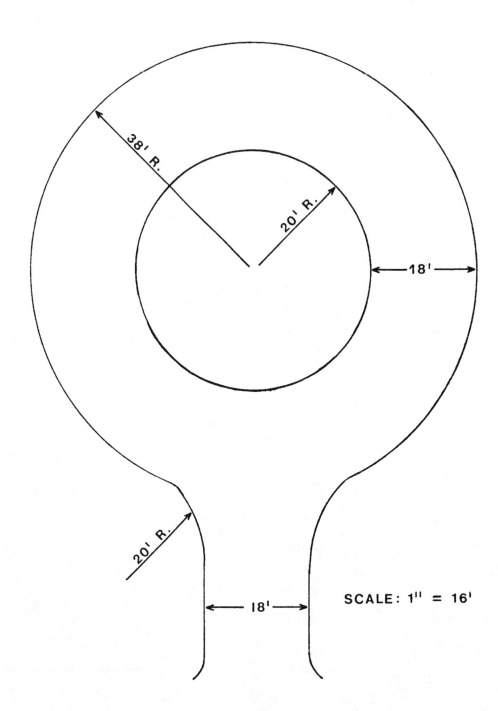

Figure 15-4 Small circular drive

Assignment

1. Study the Exercise 15 Assignment Sheets. Try to visualize the parking and/or turning needs for a two-car family.
2. Draw a guest parking and/or turning for the residence. If you are using an engineer's scale, use Assignment Sheet A and a 1:20 scale. If you are using an architect's scale, use Assignment Sheet B and a 1:16 scale.
3. Draw a front walk (4' wide) from the guest parking to the front door.
4. Complete the drive by connecting it to the street.
5. Label the areas.

Notes

Exercise 15 Assignment Sheet A

Student Name_____Date_____Score_____

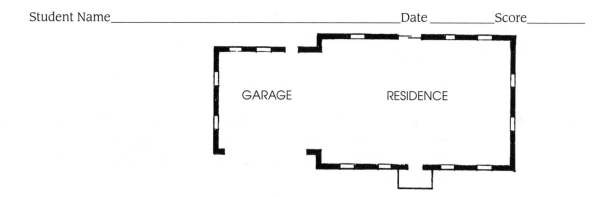

SCALE: 1" = 20'

Evaluation

Consideration	Points	Student Score	Instructor Score
Measurements are accurate	30		
Parking is appropriate	30		
Angles/radii are accurate	30		
Overall neatness	10		
Total	100		

100

Exercise 15 Assignment Sheet B

Student Name_____ Date_____ Score_____

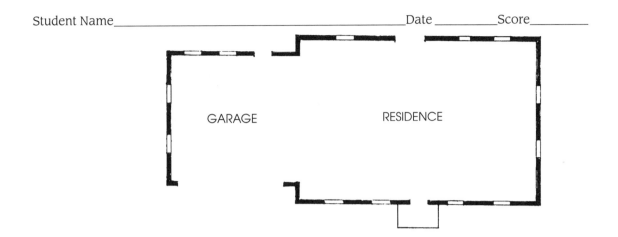

SCALE: 1" = 16'

Evaluation

Consideration	Points	Student Score	Instructor Score
Measurements are accurate	30		
Parking is appropriate	30		
Angles/radii are accurate	30		
Overall neatness	10		
Total	100		

EXERCISE 16

Designing Patios

Objective

To provide experience in designing patios for residences.

Skills

After studying this unit, you should be able to:
- understand the need for patios.
- understand the relationship between size and design to family needs.
- design an appropriate patio for a given residence, utilizing appropriate surfacing materials.
- give reasons for design and features.

Materials Needed

Drawing board and T-square
Drawing pencil (HB, F, H, or 2H)
Eraser and shield
Engineer's scale or architect's scale
Triangle
French curve(s)
Compass
Exercise 4 Sample Sheet

Introduction

A patio is an outdoor surfaced area adjoining (or convenient to) the residence that is used as an "outdoor room" by family members. The patio is located at the rear of the residence, or, occasionally, at one side. Its purpose is to provide an outdoor space for family activities during warm weather. It can be

103

used for many activities, such as barbecuing and dining, socializing, relaxing, sunbathing, and much more. A patio can provide partial or total privacy depending upon the desires of the family.

Unlike decks, patios more often are built at ground level, thus providing a kind of transition area between the residence and rear garden. The exact location along the rear of the home is dependent upon rear entrances. One or more entrances, with a minimum of steps, will greatly enhance usage by family members.

Most modern homes, unless custom built, come complete with a patio or deck. (Decks are considered in Exercise 17.) A patio is usually rectangular and made of concrete, although many different surfacing materials are available. Contractors tend to build patios that are smaller than desirable. Although a patio can be almost any size, it should be no smaller than an average room. Of course, the size can be expanded in future years. It would be helpful to take future expansion into consideration when planning the landscape.

The shape of patios vary greatly, although, as previously stated, most are rectangular. Almost any geometric shape, or combinations of shapes, can be used. In addition, curvilinear or free-form shapes are well suited to some designs. Be careful not to let shape detract from the usefulness or convenience of the patio.

Surfacing should be of permanent, easy-to-maintain material. The most commonly used surface materials are concrete, brick, and flagstone. Concrete is usually the least expensive, while flagstone is the most expensive.

The following suggestions should prove helpful in planning the patio:

1. Decide in advance the features desired, such as tables, lounge chairs, barbecue grills, etc.
2. Plan a size that will accommodate family needs.
3. Determine the most convenient location.
4. Choose a surfacing material that is durable and has a nonslip surface when wet.
5. Decide the degree of privacy desired, and plan for any shrubs, fence sections, etc., that are needed.
6. Determine the degree of bright sunlight during hours of peak usage. Plan for any shade trees or arbors that might be needed.
7. Consider night usage and whether additional lighting or insect control will be needed.

Study Figure 16-1, then complete the assignment for this exercise.

Designing Patios 105

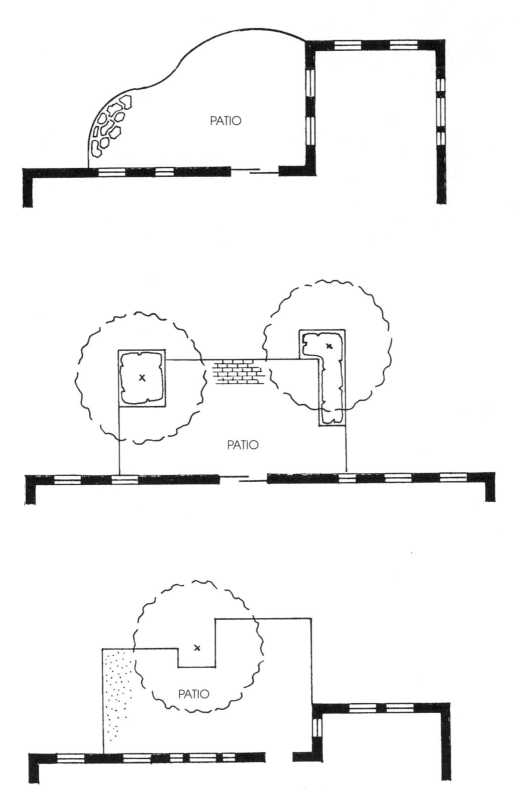

Figure 16-1 Examples of patio designs for residences

Assignment

1. Design a patio for the residence on the Exercise 16 Assignment Sheet. Use Assignment Sheet A for a 1:10 scale or Assignment Sheet B for a 1:8 scale.
2. Show any permanent features such as benches, barbecue pits, etc.
3. Using the procedures and symbols from Figures 4-1 and 4-2, indicate the surfacing.
4. Be prepared to defend the rationale (reasoning) for the size, shape, location, and any features. Present your reasons orally or in written paragraph form.

Notes

Exercise 16 Assignment Sheet A

Student Name_____Date_____Score_____

SCALE: 1" = 10'

Evaluation

Consideration	Points	Student Score	Instructor Score
Design is "feasible"	30		
Surfacing is symbolized	20		
Rationale is acceptable	30		
Labeling is accurate	20		
Total	100		

107

Exercise 16 Assignment Sheet B

Student Name_____Date_____Score_____

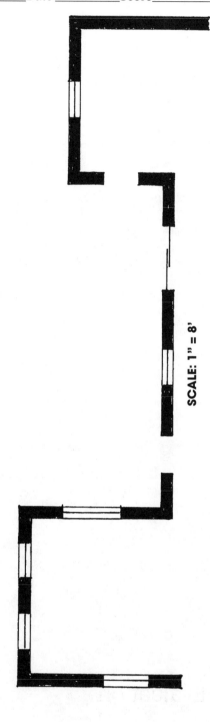

Evaluation

Consideration	Points	Student Score	Instructor Score
Design is "feasible"	30		
Surfacing is symbolized	20		
Rationale is acceptable	30		
Labeling is accurate	20		
Total	100		

108

EXERCISE 17
Designing Decks

Objective

To provide practical experience regarding balance in a residential landscape.

Skills

After studying this unit, you should be able to:
- understand the need for decks.
- understand the relationship between design and family needs.
- design a deck "floor plan" for a given residence.
- give reasons for design and any features.

Materials Needed

Drawing board and T-square
Drawing pencil (HB, F, H, or 2H)
Eraser and shield
Engineer's scale and architect's scale
Triangles
Compass

Introduction

A deck is an outdoor area adjoining the residence, that, like a patio, serves as an outdoor space for family activities. However, the deck is made of wood, and it is raised above the immediate ground level. The deck is really a structure and, as such, requires a structural support system that can range from simple to complex, depending upon the details of the design.

Structural design will not be considered in this unit. Those students desiring to learn more about the structural design of decks should consult a local building materials supply house. One or more employees can usually give advice on standard structural support specifications. Many such businesses stock brochures and booklets regarding deck support systems. Your project will involve the surface design of the deck.

A well-designed deck can be an addition of great beauty to the residence. However, one must be careful to place usefulness as the primary consideration. One advantage of a deck is that it may exist on more than one level to accommodate upper levels or sloping terrain. One disadvantage is that steps must be provided to access the rear garden. The location(s) of access steps should be well-planned.

The deck should be built of treated lumber or decay-resistant wood such as redwood. The pattern of the wooden boards can be part of the overall design and add additional beauty. Permanent benches can be added as a part of the design and constructed of the same materials. Railing is essential since the deck is raised above ground level. The verticals of the railing should be close enough to prevent young children from getting through.

Hot tubs or aboveground pools are often added, and they are great compliments to a raised deck.

Study Figure 17-1 before beginning your design. Note that the boards are shown and constitute part of the overall design.

EXERCISE 18

Understanding Balance

Objective

To provide practical experience regarding balance in a residential landscape.

Skills

After studying this unit, you should be able to:

- understand the meaning of balance in a landscape.
- understand how symmetrical and asymmetrical balance are achieved.
- demonstrate an understanding of balance by completing an assignment.

Materials Needed

Drawing board and T-square
Drawing pencil (HB, F, H, or 2H)
Eraser and shield
Engineer's scale or architect's scale
Exercise 10 Sample Sheets

Introduction

Balance in a landscape is a condition where the foliage on one side of a view is roughly equal to the foliage on the other side. Balance is often taken for granted when it exists; however, it is very noticeable or obvious when it doesn't exist. Homeowners will often shift or relocate furniture in a particular room until the right balance of items is achieved. We feel more comfortable or more at ease when things are balanced. Landscape plantings are balanced using foliage mass as a measure of balance. For instance, one small tree might balance three medium-size shrubs.

Exercise 18

Balance may be divided into symmetrical and asymmetrical. Symmetrical balance is achieved when the plantings on one side of a view are the exact same as on the other. This type of balance is used when the residence or yard is exactly symmetrical. Symmetrical balance tends to be more formal, so caution (restraint) should be exercised.

Asymmetrical balance is achieved when the plantings are not a mirror-image on each side, but the plantings have somewhat equal foliage weight. This type of balance is less formal or stiff, and it is the most desirable for the majority of residences.

A more advanced approach to balance, for the experienced designer, considers color, form, and texture. Color would be off-balance if all the plants on one side were red-leaved and all plants on the other side were green-leaved. The same would be true for shape of the plant (form) and size of leaves (texture). The Exercise 18 Assignment Sheet will consider just foliage weight in achieving balance.

Study Figure 18-1 before completing the assignment.

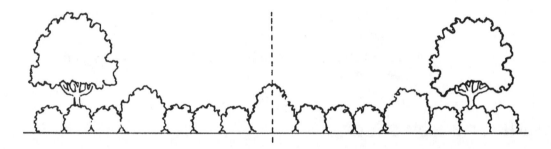

SYMMETRICAL BALANCE

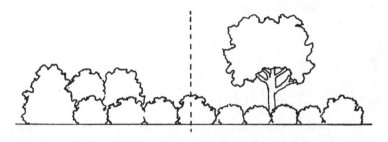

ASYMMETRICAL BALANCE

Figure 18-1 Symmetrical and asymmetrical balance

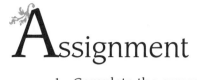 Assignment

1. Complete the symmetrical balance part of the assignment sheet. It has been started for you. Fill in a mass of plants along the entire line to achieve symmetrical balance.
2. Complete the assignment for asymmetrical balance using the same procedure as above.
3. Both the sample sheet and assignment sheet are drawn on a 1:10 scale.

 Notes

Exercise 18 Assignment Sheet

Student Name_____Date_____Score_____

Fill the blank space to achieve <u>symmetrical balance</u>.

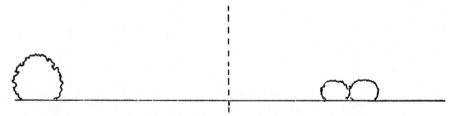

Fill the blank space to achieve <u>asymmetrical balance</u>.

Evaluation

Consideration	Points	Student Score	Instructor Score
Balanced is achieved	50		
Plantings are o scale	30		
Overall neatness	20		
Total	100		

EXERCISE 19
Designing Curvilinear Gardens

Objective

To provide practical experience in designing a curvilinear border.

Skills

After studying this unit, you should be able to:
- understand the objective(s) of curvilinear design in a landscape.
- understand the terms informal, massing, variety, texture, and repetition as applied to landscape design.
- complete a design for a backyard border using curvilinear design.

Materials Needed

Drawing board and T-square
Drawing pencil (HB, F, H, or 2H)
Eraser and shield
Engineer's scale or architect's scale
French curves
Circle template
Figures 5-1, 5-2, and 6-1
Figure 12-1

Introduction

Just as people vary in their "tastes" for homes, automobiles, and food, so their tastes vary for plants and their arrangement. For example, one extreme is the woodland retreat with no added landscape plantings. On the other extreme is the formal geometric pattern found in some European estates (i.e., an Elizabethan garden). Somewhere between the two extremes is the design appropriate for most modern residential properties.

The most popular design in residential landscaping is the **curvilinear** design. This concept uses long, sweeping curves to outline planting beds. This approach minimizes abrupt corners and straight lines that are often characteristic of the more formal geometric designs (considered in Exercise 20). The curvilinear pattern is considered more "informal" and invites a more relaxed mood. This is especially important for the family that places great emphasis on a casual lifestyle when at home.

Review Figure 12-1 regarding a curvilinear design for a small rear garden. Notice that the curving lines are long, sweeping lines. Short, choppy lines are difficult to mow, and they give a confused or overdone look. These lines are formed where the turfgrass meets the mulch in the planting beds. These lines should be formed by tools (equipment) and maintained prominently for maximum attractiveness. The maintenance of these lines may be the single most important consideration in a beautiful landscape.

Once the general outline is decided, begin designing the plants for the areas. The following guidelines should be considered:

1. **Massing**–This concept has two considerations. First, it means that plants of a kind should be grouped together, rather than alternating. Second, it means that plants are placed close enough to look like a mass without overcrowding.

2. **Variety**–Always use several varieties of plants to insure that the design will be interesting. There should be variations in time of bloom or leaf color, and mature height.

3. **Texture**–Texture is most often associated with leaf size. Do not use all coarse textured or all fine-textured plants. Variation will add interest.

4. **Repetition**–Repeat some plants throughout the design. For example, some of the plants used on one side of the yard should be repeated on the other. In symmetrical designs, the numbers will be the same. In asymmetrical designs, it is not necessary to use the same number.

Review Exercise 12 for examples of the above, then complete the assignment for this exercise.

ssignment

1. Study Figure 19-1 as an example of a curvilinear design for a small rear garden.
2. IMPORTANT: Turn to the Introduction section of Exercise 24 and study the "three magic keys." Use these "keys" in your design.
3. Complete the Exercise 19 Assignment Sheet by drafting a curvilinear design for the rear garden illustrated. Be sure to balance the design. Use Assignment Sheet A for a 1:10 scale or Assignment Sheet B for a 1:8 scale.

otes

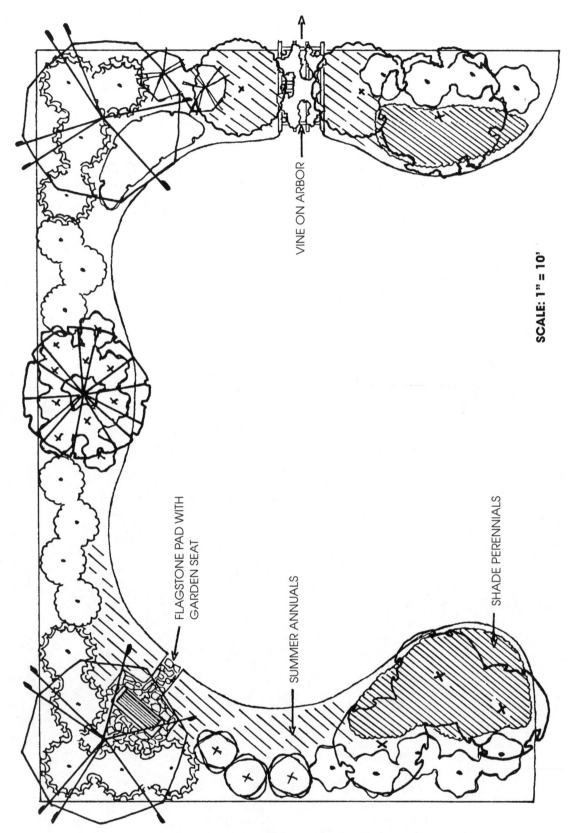

Figure 19-1 Example of curvilinear design

Exercise 19 Assignment Sheet A

Student Name_____Date_____Score_____

5' BRICK WALL

ACCESS GATE

SCALE: 1" =10'

Evaluation

Consideration	Points	Student Score	Instructor Score
Lines are curvilinear	30		
Choppy lines were avoided	20		
Plants are massed	20		
Variety is planned	10		
Repetition is utilized	10		
Drawing is to 1:10 scale	10		
Total	100		

123

Exercise 19 Assignment Sheet B

Student Name_____Date_____Score_____

(Use Evaluation on Assignment Sheet A.)

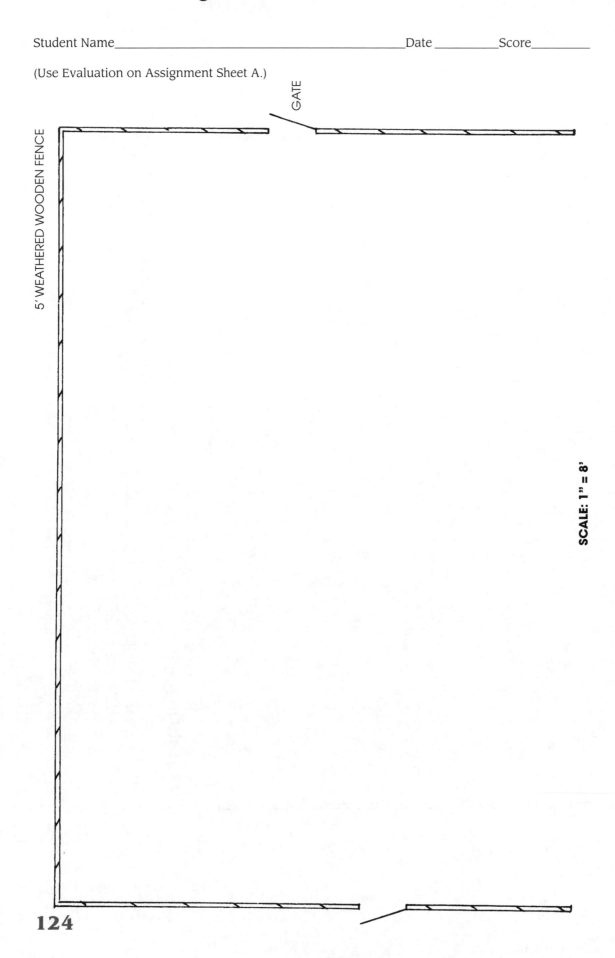

EXERCISE 20
Designing Geometric Gardens

Objective

To provide practical experience in designing a geometric garden.

Skills

After studying this unit, you should be able to:

* understand the meaning of geometric design.
* complete a design for a border using geometric design.

Materials Needed

Drawing and T-square
Drawing pencil (HB, F, H, or 2H)
Eraser and shield
Engineer's scale or architect's scale
Triangles
Circle template
Symbol sheets for Exercises 5 and 6

Introduction

Geometric designs utilize geometric shapes in laying out bed areas, lawn areas, and other areas of the landscape. The most commonly used shape is the rectangle (rectilinear), although circles, squares, triangles, or combinations of shapes are used. Such designs are distinctly more formal, and they are often associated with a more formal lifestyle. Some of the most elaborate geometric designs are found on the grounds of castles and palaces throughout the world. These designs are quite descriptively referred to as formal gardens.

Geometric designs for most residences, while formal, are usually much less elaborate in design when compared to a palace or formal estate. For the homeowner, such designs utilize all the features of curvilinear design, except that the "lines" are geometric rather than curving. Balance, repetition, variety, etc., are just as important. Geometric gardens can be very attractive, but they sometimes require more time in design in order to achieve maximum beauty.

There are two important considerations in designing geometric gardens:

1. Geometric gardens look best on flat or level ground. Where there is slope, it is often desirable to divide the yard into two or more flat areas with convenient access from one level to another.

2. With the exception of circles, geometric designs should utilize parallel lines. These lines should be parallel to some feature such as a wall or the "line" of a deck. Sometimes, two or more patterns of parallel lines are used in the same design.

Study Figure 20-1; then, complete the assignment for this exercise.

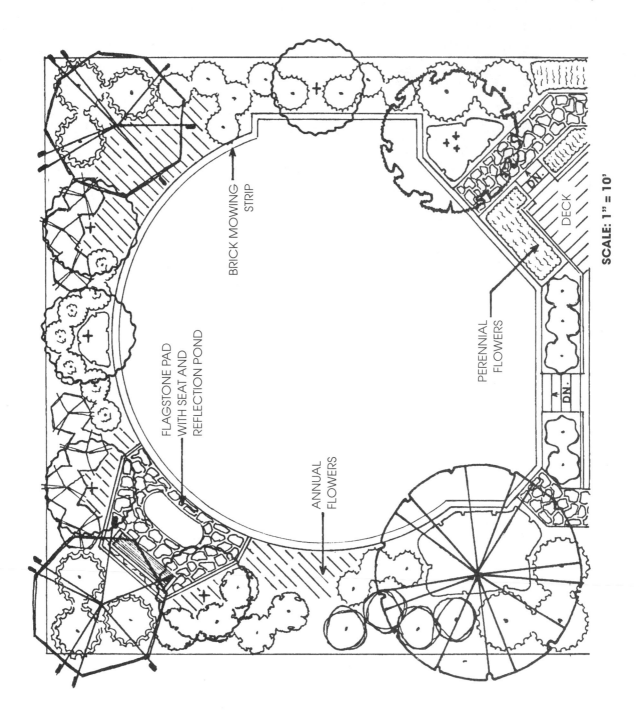

Figure 20-1 Example of a geometric designed garden

ssignment

1. Study Figure 20-1.
2. Complete the Exercise 20 Assignment Sheet by drafting a geometric design for the area illustrated. Utilize all previous relevant concepts in the design. Use Assignment Sheet A for a 1:10 scale or Assignment Sheet B for a 1:8 scale.

otes

Exercise 20 Assignment Sheet A

Student Name_____Date_____Score_____

SCALE: 1" = 10'

Evaluation

Consideration	Points	Student Score	Instructor Score
Lines are geometric	30		
Parallel lines are used	20		
Appropriate concepts applied	20		
Drawing to 1:10 scale	20		
Overall neatness	10		
Total	100		

Exercise 20 Assignment Sheet B

Student Name_____ Date_____ Score_____

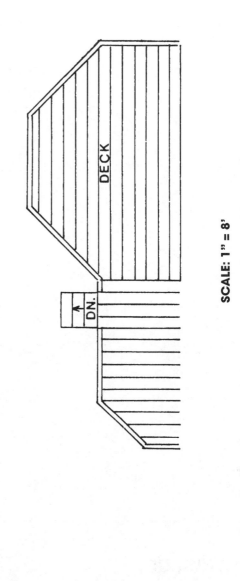

Evaluation

Consideration	Points	Student Score	Instructor Score
Lines are geometric	30		
Parallel lines are used	20		
Appropriate concepts applied	20		
Drawing to 1:8 scale	20		
Overall neatness	10		
Total	100		

Understanding Focalization

Objective

To provide experience in designing "focal points" for garden areas.

Skills

After studying this unit, you should be able to:
- understand the concept of focalization (focal points).
- develop and/or identify a "focal point" in a garden design.

Materials Needed

Drawing board and T-square
Drawing pencil (HB, F, H, or 2H)
Eraser and shield
Engineer's scale or architect's scale
Circle template
Triangles
Rear garden design for Exercise 20

Introduction

In developing the residential design, it is often necessary to have well-planned areas that immediately catch the eye when entering an area of the garden. Such areas are called "focal points," and the concept of planning their use in the landscape is called **focalization.**

Focal points, while important, should be kept to a minimum to avoid competition between the focal points themselves. In the front yard or public area, the primary focal point is the front door. Less dramatic attention can be drawn to the guest parking area or the entrance to the front walk. In the rear garden or private living area, a focal point should provide an attractive view from the patio or deck for the enjoyment of family members and guests (see Figure 12-1).

Focalization can utilize specimen plants that are very noticeable, natural features such as streams, and/or man-made features such as statuary (see Figure 12-2). When plants alone are used, it is possible to shift points throughout the seasons. For example, a flowering tree might serve as the focal point for an attractive area of the garden during spring, then the focus might shift to a grouping of flowering shrubs during summer. Fall color might shift the focus in fall. Where different focal points occur, it is important that the area attracting attention be neat or worthy of the attention. In addition, it is important that no two areas compete at the same time. One technique of merit is to have plants of varying flowering times in the same bed area. For example, the spring flowering tree might have summer flowering plants growing under it. This will keep the attention on your favorite area.

Of all the techniques involved in the development of the plan, the concept of focalization requires as much thought in design as any other concept. Study Figures 21-1 and 21-2, then complete the Exercise 21 Assignment.

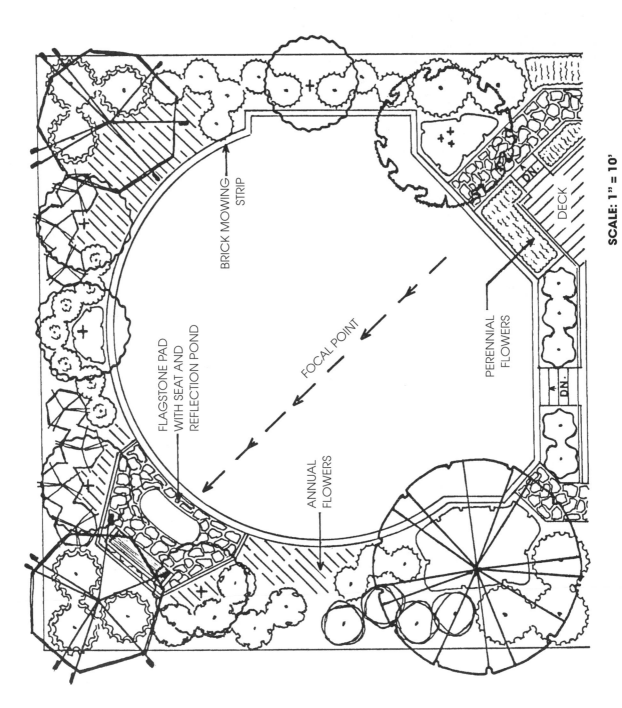

Figure 21-1 Focal point in a rear garden or private living area

Exercise 21

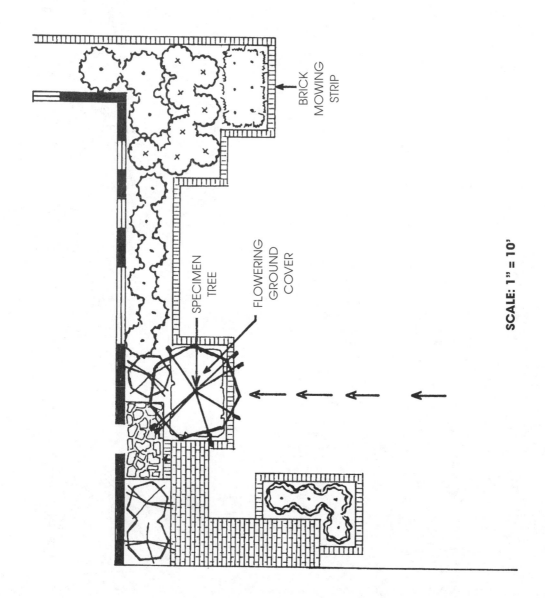

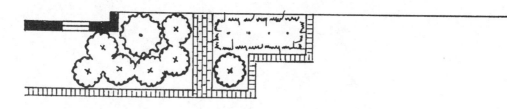

Figure 21-2 Focalization utilizing plants and man-made features

Assignment

1. Using the Exercise 20 Assignment Sheet where you designed a rear garden, modify the plan to include a focal point. You may need to erase some features to make space for the focal point. Draw arrows from the logical viewing area of the residence to the focal point itself. If you have already included a focal point, you need only to draw the arrows. (Refer to Figure 21-1.)
2. Study Figure 21-2, then draft a focal point for a front entry using Assignment Sheet A for a 1:10 scale or Assignment Sheet B for a 1:8 scale.

Notes

Exercise 21 Assignment Sheet A

Student Name_____Date_____Score_____

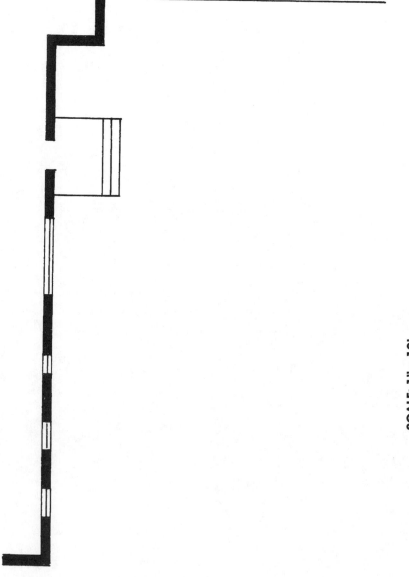

SCALE: 1" = 10'

Evaluation

Consideration	Points	Student Score	Instructor Score
Focal points appropriate	40		
Features are labeled	30		
Exact scale utilized	20		
Overall neatness	10		
Total	100		

Exercise 21 Assignment Sheet B

Student Name_____Date_____Score_____

SCALE: 1" = 8'

Evaluation

Consideration	Points	Student Score	Instructor Score
Focal points appropriate	40		
Features are labeled	30		
Exact scale utilized	20		
Overall neatness	10		
Total	100		

Notice to the Student:

The remaining exercises will constitute your final landscape plan. All remaining exercises will be performed on your sheet of drafting vellum. Use a minimum size of 17" x 22" if using a 1:10 scale or a minimum size of 24" x 36" if using a 1:8 scale.

Organizing/Beginning the Planting Plan

EXERCISE 22

Objective

To provide experience in "laying out" the complete residential landscape planting plan on a sheet of drafting vellum.

Skills

After studying this unit, you should be able to:

- ✤ determine the size of paper needed to draft the planting plan.
- ✤ determine the location of, and the space needed for, the lot, plant list, and title block.
- ✤ prepare the vellum for drafting, then draft the property for a residence onto the vellum.

Materials Needed

Drawing board and T-square
Drawing pencil (HB, F, H, 2H)
Engineer's scale and architect's scale
Eraser and shield
One sheet of drafting vellum–17" x 22" (1:10 scale) or 24" x 36" (1:8 scale).
Drafting tape (not masking tape!)

Introduction

When drafting a landscape plan, it is desirable to have all information for the property, or a designated area of the property, on one sheet of vellum. The information should include the actual design, a plant list, and a title block. The property size and plant list must be considered when deciding the size of paper needed for the project. It is desirable to use the smallest paper size that will provide adequate space for the project, without overcrowding.

To get started, always have the longest side parallel to the top and bottom edges of the drawing board. Drafting vellum is rectangular in shape and comes in various sizes. The standard sizes are 17" x 22", 18" x 24", 24" x 36", 30" x 42", and 36" x 48". If paper larger than 36" x 48" is needed, one should divide the property into areas, such as Area "A", Area "B", etc.; or use a smaller scale, such as 1:20. Scales smaller than 1:16 or 1:20 are not recommended for landscape drafting.

Residential properties are almost always somewhat rectangular in shape; however, many shapes exist. In drafting the property, always draft the front yard adjacent to the bottom edge of the paper and the back yard near the top of the paper. Also, the property should be drafted on the left side of your vellum, with the plant list and title block on the right.

In using the scale, it is easy to determine the size of paper needed. Always allow a minimum of one-half inch space between the border and any edge of the property. A lot 120' wide and 180' deep would not fit on a 17" x 22" sheet of vellum; therefore, a 24" x 36" sheet would be needed. A lot 120' wide and 150' deep would fit on a 17" x 22" paper.

To draft the property, determine the size of paper needed and draw a one-half inch border on all four sides. Next, plan for equal "empty" spaces between the front and rear property lines. Last, plan for equal "empty" spaces between the left border and left property line; between the right property line and plant list; and between the plant list and right border of the paper. Allow around 10" of space for the plant list. (Plant lists and title blocks will be considered in future exercises.)

See Figures 22-1A and B for examples.

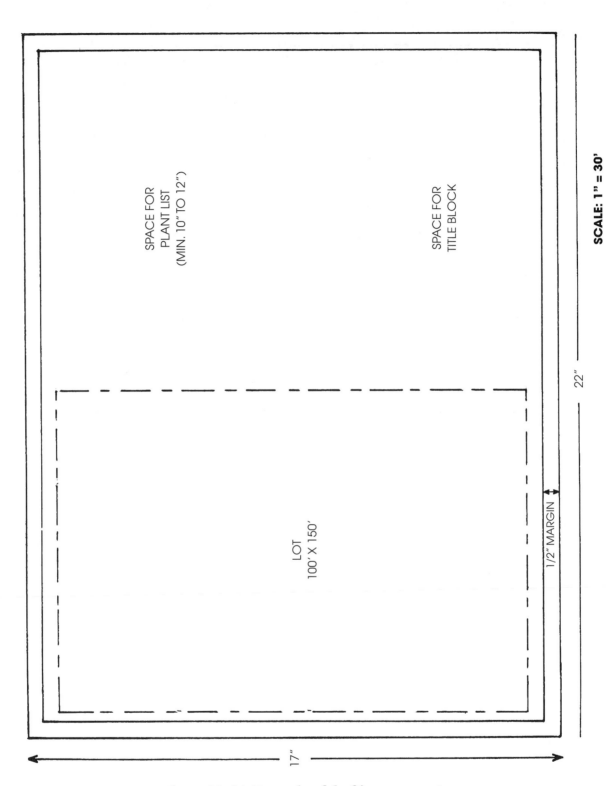

Figure 22-1A Example of drafting a property

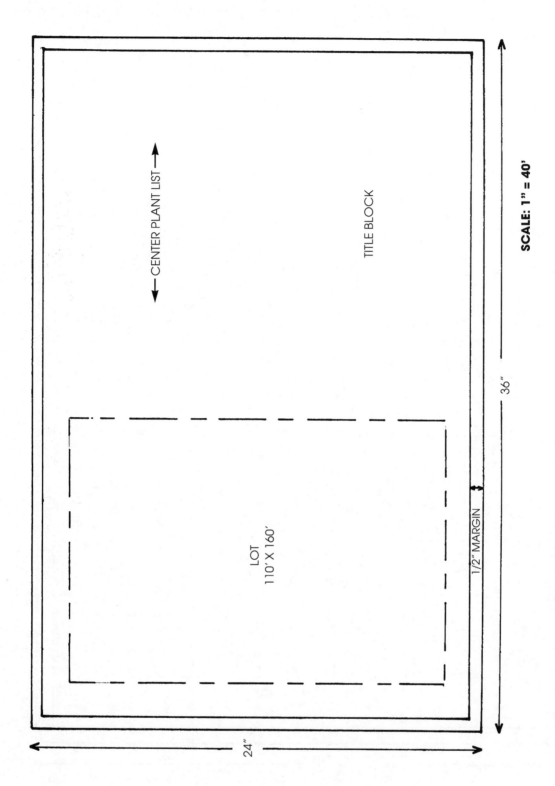

Figure 22-1B Example of drafting a property

Exercise 23 Evaluation

Student Name_____ Date_____ Score_____

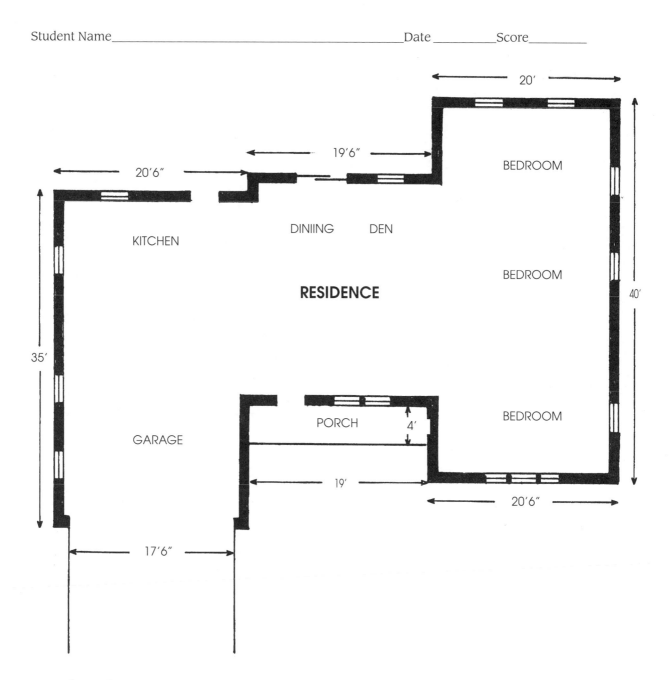

Evaluation

Consideration	Points	Student Score	Instructor Score
House located appropriately	30		
Walks, drive, etc., are acceptable	30		
Exact scale maintained	20		
Overall neatness	20		
Total	100		

147

Developing the Design

Objective

To assist students in the development of the "total design" for a residential landscape.

Skills

After studying this unit, you should be able to:

- utilize previously acquired skills in the development and drafting of a landscape for a residential property.
- organize the yard into functional areas.
- label items as per previous instruction.
- implement the "three magic keys."
- utilize a checklist in preevaluation.

Materials Needed

Drawing board and T-square
Drawing pencil (HB, F, H, or 2H)
Engineer's scale or architect's scale
Eraser and shield
Triangles
Compass
French curves
Circle template
Sheet of vellum used in Exercise 23
Drafting tape

Introduction

You are about to begin the most time-consuming exercise–devising the arrangement of plants in the total design. Care should be used in placing plants, as this will determine more than any one factor, the quality of the design. You may need to experiment, so do not hesitate to use your pencil and eraser in making desired improvements. You should begin with the foundation planting for the residence.

In designing the total yard plan, there are "three magic keys" to an attractive design:

1. Landscape the borders of the property. This is especially important in the rear garden.
2. Leave open areas of lawn. These areas will vary greatly in size from one garden to another, but they are present, to some degree, in the very best landscapes.
3. Use curving lines in the borders (except for geometric designs). The curving design (curvilinear) is far more popular, because it gives a more natural look.

Refer to Figure 12-1 for examples of these "keys."

A Landscape Checklist has been included for your use in evaluating the design phase of your plan. The following is a brief explanation of checklist items. Some items will be a review of previously considered materials.

I. Foundation Planting
   A. Use taller plants on corners. Medium-size shrubs should be used for one-story homes; large shrubs for two-story or taller.
   B. Use dwarf shrubs or ground covers under windows 4' or less above ground level.
   C. Enhance the entry by using noticeable or different plants.
   D. Balance the planting with equal "foliage mass." Remember: Symmetrical structures must have the same number and species of plants on each side of the front door.
   E. Repeat some of the same plants on each end, even if it is an asymmetrical landscape.
   F. Variety should be used in selecting plants. This includes texture (leaf size).
   G. Massing of plants will give a more pleasing effect. Plants should be drawn to mature size and allowed to touch, or almost touch, adjacent plants.

II. Overall Design of Property
   A. Balance—The entire foliage mass should appear in balance.
   B. Repetition—Same as for foundation plants.

C. Variety and texture—Same as for foundation plants.

D. Massing—Same as for foundation plants.

E. Simplicity—Absence of inconveniences, such as unnecessary curves in walks, fences without gates, etc. (Repeating some plants can be included here.)

F. Circulation—Ease in getting around in the garden. Can you get a truck in the rear garden, if necessary?

G. Areas defined—Yard is divided into functional areas: public, private, utility, and, if needed, a play area for children.

H. Focal points—Areas that draw attention, especially in the rear garden and front entry of the residence.

I. Avoid straight lines of plants, if possible.

J. Provide guest parking.

K. Borders of property should be landscaped.

L. Open areas of lawn should be planned.

M. Borders should be curving (except in geometric designs).

III. Drafting Skills

A. Use exact scale.

B. Letter items that need labeling. Be neat.

C. Strive for overall neatness.

D. Lines should be dark and smooth.

E. Different plant symbols are used for different plant species.

Assignment

1. Using the house and property drafted in Exercise 23, lightly sketch bubble diagrams of areas and features. Lines can be erased as you develop permanent lines.

2. Draft the entire plan, using reference material from this exercise and previous exercises.

3. Maintain exact scale for every feature.

4. Keep a separate list of plants used; you will learn to select plants and label plants in future exercises.

5. Take your time and strive for neatness.

6. Be creative, but don't "get carried away."

Landscape Checklist

Use this form for preevaluation of your design. Place an "X" in the block beside an item that is acceptable, and correct any deficiencies before completing the exercise Evaluation.

Foundation Planting

- ❏ Appropriate size plants were used on the corners of the house.
- ❏ Dwarf plants or ground covers were used under the windows.
- ❏ Entry to the residence has been enhanced.
- ❏ Foliage mass appears balanced.
- ❏ Plants were repeated in the design of the foundation planting.
- ❏ Variety in plants and texture was used.
- ❏ Plants were massed.

Overall Design of Property

- ❏ Overall design of the property appears balanced.
- ❏ Plants were repeated throughout the design.
- ❏ Variety in plants and textures was used.
- ❏ Simplicity is evident.
- ❏ Circulation provides ease in movement throughout the garden.
- ❏ Areas are well defined and appear organized.
- ❏ Focal points exist in both the front and rear gardens.
- ❏ Straight lines have been avoided wherever possible.
- ❏ Guest parking has been provided and is appropriate.
- ❏ Plantings have been designed into the borders of the yard.
- ❏ There are open areas of lawn planned.
- ❏ Borders are curving (or geometrically designed).

Drafting Skills

- ❏ Exact scale was used throughout the design.
- ❏ Lettering is both complete and neatly drafted.
- ❏ Lines are dark, smooth, and consistent.
- ❏ Overall neatness was maintained throughout the entire design.

Exercise 24 Evaluation

Student Name_____Date_____Score_____

Evaluation

Consideration	Points	Student Score	Instructor Score
Foundation planting acceptable	30		
Overall design acceptable	30		
Lettering is adequate and neat	10		
Exact scale used	20		
Overall neatness	10		
Total	100		

Exercise 25: Selecting Plants

Objective

To provide students with experience in plant selection skills.

Skills

After studying this unit, you should be able to:

- understand size variations or size groups in mature plants.
- understand uses of various groups.
- locate sources of information on plants.
- prepare an informal list of plants.

Materials Needed

Notebook paper and pencil or pen
Drawings from Exercises 23 and 24 (reference purposes)
Various books*, magazines, booklets/brochures, and nursery catalogs containing specific information on landscape plants. Usually the local extension service can provide information on the plants for your area. Ask your instructor to provide assistance in locating pictures or specific information on plants. Do not worry over this exercise, since design is the primary objective for this unit.

NOTE: A great reference for use in selecting plants for your plan is Landscape Plants–Their Identification, Culture, and Use *by Ferrell Bridwell, and published by Delmar Publishers.*

Introduction

In the selection of landscape plants, it is necessary to group plants according to the sizes they attain at maturity. For example, dwarf shrubs or large trees are such groups, and each contains many fine plants suitable for landscape purposes. One of the first questions people ask is, "How big does it get?" Information on specific plants should include:

1. Region(s) in which it grows.
2. Size–height and width.
3. Information on leaves, flowers, and fruit.
4. Light requirements, such as sun or shade.
5. Habit of growth or shape of the plant.
6. Evergreen or deciduous (loses leaves in fall).

The reference previously listed *(Landscape Plants)* has simplied these items for quick reference and ease in selecting plants.

The following will describe size groups according to height, as well as some possible uses for plants in each group.

Low-Growing Ground Covers—0 to 2'
Ground covers are used as a substitute for turfgrasses where mowing might be difficult, under low-branched trees, shady areas where grass might not grow, and to add variety to the landscape.

Vines—Size varies
Vines can be trained on walls, arbors, trellises, and fences. Vines are good substitutes for shrubs where space is limited, and they add interest.

Dwarf Shrubs—Less than 4'
Dwarf shrubs are used in foundation plantings, especially under windows. They are often used as low-growing hedges or as a substitute for turfgrasses or ground covers. They are best used in masses in "bed" areas in the landscape.

Medium-Size Shrubs—4' to 6'
Medium shrubs are used on corners of one-story structures, unclipped hedges, border plants, and as background for smaller plants.

Large Shrubs—6' to 12'+
Large shrubs are used as corner plants for taller structures, windbreaks and hedges, and as tree-formed specimens.

Small Trees—Less than 40'
Small trees are used for shade or accent and to give variety. They look nice in groupings and do well in smaller areas where a large tree would not be appropriate.

Large Trees—Over 40'
Large trees are used as screens, specimen plants (in many cases), and for shade.

Speciality Plants—Great variation
These include ornamental grasses, bamboo, palms, fruit trees, banana trees, etc. These plants have a variety of functions in landscapes. You should study these plants carefully before using them in designs.

Assignment

1. Prepare an informal list of plant names for the plants in your plan (use Figure 25-1 as an example). You will need both common names and botanical names.
2. This informal list will be drafted onto your plan in the next exercise.
3. A good landscape plan should contain a minimum of 20 different species of plants. Be sure to select 20 or more.

A   2   Red Maple - *Acer rubrum* "October Glory"

B   1   River Birch - *Betula nigra*

C   2   Eastern Redbud - *Cercis canadensis*

D   1   Yoshino Cherry - *Prunus yedoensis*

E   3   Flowering Dogwood - *Cornus florida* "Cloud Nine"

F   9   Hybrid Rhododendron - *Rhododendron* x "Roseum Elegans"

G   7   Manhattan Euonymus - *Euonymus kiautschovicus* "Manhattan"

H   2   English Holly - *Ilex aquifolium*

I   3   Pampas Grass - *Cortaderia selloana*

J   9   Mountain Laurel - *Kalmia latifolia*

K   14   Kurume Azalea - *Azalea obtusum* "Hino-Crimson"

L   2   Hybrid Clematis - *Clematis* x *jackmanii*

M   9   Helleri Holly - *Ilex crenata* "Helleri"

N   193 Japanese Spurge - *Pachysandra terminalis*

Figure 25-1 Informal list of plants

Exercise 25 Assignment Sheet

Student Name_____ Date _____ Score_____

Evaluation

Consideration	Points	Student Score	Instructor Score
List contains common and botanical names	40		
Plants are suitable for your region	40		
List contains 20 or more species	20		
Total	100		

Exercise 26: Labeling Plants on a Plan

Objective

To provide experience in labeling the plants on a landscape plan.

Skills

After studying this unit, you should be able to:
- understand the need for labeling plants.
- "key" or code different plants on a landscape plan.

Materials Needed

Drawing board and T-square
Drawing pencil (HB, F, H, or 2H)
Eraser and shield
Engineer's scale or architect's scale
Triangles
Drawing from Exercises 23, 24, and 25
Drafting tape
"Informal" plant list from Exercise 25

Introduction

In order for a landscape plan to be usable, one must correctly label all plants in such a way that others can "read" the plan. One system is to write names on the plan near the plant and draft a line from the name to the plant. On complete plans, this system can become confusing with a mass of names "cluttering" the plan.

A proven method is to use "keys" or codes to identify the plants on the plan. These codes can then be matched to codes in the actual plant list to find the name of the plant (see Figure 26-1). Such codes use letters of the alphabet to "key" the plants and numbers to give the quantity of plants in the group or area.

One approach is to start coding in the front yard and "work" your way to the rear. Some instructors prefer to list trees first, then shrubs. Give the first plant(s) you identify the code of A. If you have only one of "A", then label it as A-1. If there is a group of plants, for example, five, of the same species, then write A-5. Draw a line to the plant or to the group. The next plant you code will be identified as B. If your list contains more than twenty-six different plants, then, after Z has been used, start labeling AA, BB, etc. The codes should be parallel to the bottom edge of the sheet and written in $\frac{1}{10}$" or $\frac{1}{8}$" capital letters.

Study Figure 26-1. Have your instructor provide additional explanations, if needed.

Assignment

"Key" or code the plants located on your plan. Record the codes on your "informal" list for future reference and use (refer to Exercise 25). The actual plant list will be drafted onto the plan sheet in the next exercise.

Notes

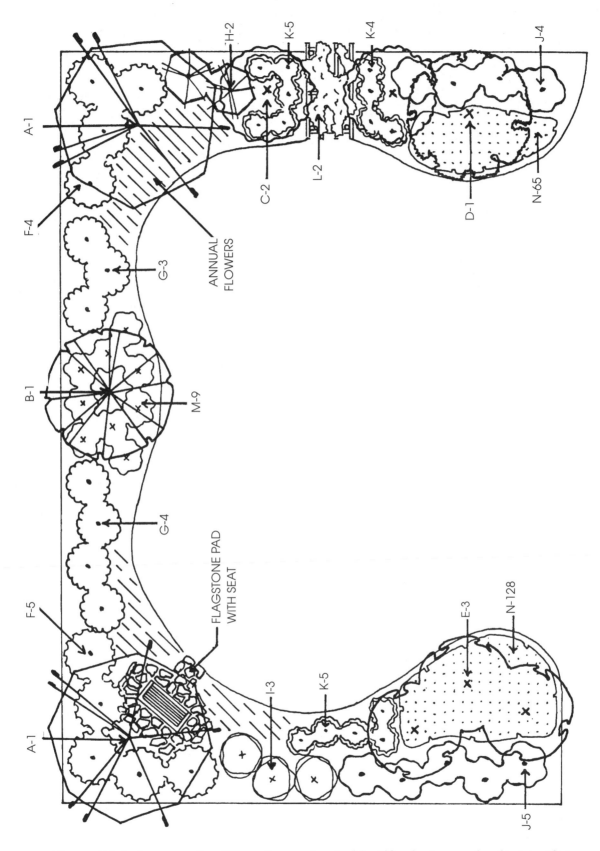

Figure 26-1 An example of "keys" or codes to identify plants on a landscape plan

Exercise 26 Evaluation

Student Name_____Date _____Score_____

Evaluation

Consideration	Points	Student Score	Instructor Score
All plants, or groups, are coded	60		
Lettering is parallel to bottom edge	20		
Lines connect codes to plants	20		
Total	100		

Exercise 27: Preparing a Finished Plant List

Objective

To assist students in drafting the final plant list onto the plan.

Skills

After studying this unit, you should be able to:

- draft the plant list onto the plan, providing appropriate information.
- use guidelines to properly align codes, numbers, and names.
- understand why botanical names are placed on the plan.

Materials Needed

Drawing board and T-square
Drawing pencil (HB, F, H, or 2H)
Eraser and shield
Engineer's scale or architect's scale
Informal plant list from Exercises 25 and 26
Drawing from Exercises 23, 24, 25, and 26

Introduction

An accurate plant list can provide a quick reference of the plants in the plan and the quantity of plants for each species. If accurate, the plant list alone can be used to purchase the plants for the landscape. You will need to provide four kinds of information for the plant list–codes, quantity of each plant, common names, and botanical names. Some plans go a step further and provide the size to be purchased.

1. Key—As previously discussed, you will use letters of the alphabet to code the different plants.
2. Quantity (abbreviated "Quan.")—Add all quantities for the same code, and list the total quantity under the heading, Quan. For

instance, you might have 3 of plant A in one area of the yard and 5 of plant A in another area. The Quan. for plant code A would then be 8 (3 + 5). See the smple sheet. Tally the total for each different code on your plan.

3. Common Name—This is the English name used for a particular plant in your area, such as Sugar Maple, White Oak, etc. Common names may vary from one area to another.

4. Botanical Name—The botanical name, or scientific name for a plant, is the Latin name assigned a plant. There are usually two words used. The first is the genus name, and the second is the species name. A third name is sometimes used, and it may be a variety name (Latin) or cultivar (any language and placed in single quotes). The botanical name is the only accurate name, since no two plants have the exact same botanical name.

In placing the information on the plant list, it will be necessary to use horizontal guidelines to provide the row of information for each plant. In addition, vertical guidelines will help align the information vertically in columns. Always leave a spacing equal to, or greater than, the height of the letters used.

Study Figure 27-1 before beginning the assignment.

Assignment

1. Using Figure 27-1 as a guide, prepare your plant list on the space reserved on your drawing. Use letter sizes that are approximately the sizes as shown on the sample sheet, allowing slight variations to fit the scale you are using.
2. Use both horizontal and vertical guidelines, as illustrated.
3. Be sure to include all plants on the design for the entire property.

PLANT LIST

KEY	QUAN.	COMMON NAME	BOTANICAL NAME
A	2	RED MAPLE	ACER RUBRUM "OCTOBER GLORY"
B	1	RIVER BIRCH	BETULA NIGRA
C	2	EASTERN REDBUD	CERCIS CANADENSIS
D	1	YOSHINO CHERRY	PRUNUS YEDOENSIS
E	3	FLOWERING DOGWOOD	CORNUS FLORIDA "CLOUD NINE"
F	9	HYBRID RHODODENDRON	RHODODENDRON X "ROSEUM ELEGANS"
G	7	MANHATTAN EUONYMUS	EUONYMUS KIAUTSCHOVICUS "MANHATTAN"
H	2	ENGLISH HOLLY	ILEX AQUIFOLIUM
I	3	PAMPAS GRASS	CORTADERIA SELLOANA
J	9	MOUNTAIN LAUREL	KALMIA LATIFOLIA
K	14	KURUME AZALEA	AZALEA OBTUSUM "HINO-CRIMSON"
L	2	HYBRID CLEMATIS	CLEMATIS X JACKMANII
M	9	HELLERI HOLLY	ILEX CRENATA "HELLERI"
N	193	JAPANESE SURGE	PACHYSANDRA TERMINALIS

Figure 27-1 Plant list

Exercise 27 Evaluation

Student Name_____Date_____Score_____

Evaluation

Consideration	Points	Student Score	Instructor Score
List is aligned horizontally and vertically	30		
All plants are included	40		
Scale is accurate	20		
List is neat	10		
Total	100		

EXERCISE 28
Preparing a Title Block

Objective

To guide students in the development of a symmetrical title block for the completed landscape plan.

Skills

After studying this unit, you should be able to:

- draft a symmetrical title block for a residential landscape plan.
- include all needed information in the title block.
- center each line on a central, vertical guideline.

Materials Needed

Drawing board and T-square
Drawing pencil (HB, F, H, or 2H)
Eraser and shield
Engineer's scale or architect's scale
Drawing from Exercises 23, 24, 25, 26, and 27

Introduction

The title block is the final step in the preparation of the landscape plan. It is one of the first things noticed, so it should be very neat, and it should contain all necessary information. The minimum items of information that should be included are the words "Landscape Plan for," the client's name, month and year developed, designer's name, and the scale used. In some cases it is desirable to include the drawing number, client's address, designer's company or address, and the page number if more than one page is used.

In preparing the title block, the client's name should be drafted in larger letters than other lettering. After all, who's paying for the design? Designers might desire to emphasize their names, but they should still be drafted in smaller letters than the clients' names.

The title block should be centered in the remaining space on the lower right portion of the drawing. It is unnecessary to enclose the information in a block or box, as implied. However, some vellum comes with a true block, preprinted and allowing space for all necessary information. In fact, some businesses order custom-designed vellum which contains custom-made title blocks with preprinted company name or logo.

In the absence of preprinted title blocks, the most visually pleasing title block is the symmetrical title block. To prepare a symmetrical title block, it is necessary to use several horizontal guidelines and one vertical guideline. Basically, each horizontal line is centered on the vertical line (see Figure 28-1).

One helpful technique is to draft the horizontal lines and wording on a separate sheet, slide the sheet under your drawing, center it on the vertical line, and trace it. However, it is possible to count the letters and spaces on a line and divide by two. This will help you to locate the center letter or space on a line. Remember: This is your last chance to make the design look really good. **Go for it!**

Study Figure 28-1, and use the letter sizes and spaces as indicated.

```
LANDSCAPE PLAN
       FOR
THE SMITH RESIDENCE
123 OAK STREET, SOMETOWN, USA
DRAWN BY: SUSIE QUE, LANDSCAPE DESIGNER
APRIL 199X          SCALE: 1" = 10'
```

Figure 28-1 Example of a symmetrical title block

Assignment

1. Prepare a symmetrical title block for your plan. "Make up" your own client name, and list your name as the designer.
2. This is the last exercise! I'm sure you did a great job. Good luck in the future.

Notes

Exercise 28 Evaluation

Student Name_____ Date_____ Score_____

Evaluation

Consideration	Points	Student Score	Instructor Score
Title block is symmetrical	60		
Lettering is accurate and neat	40		
Total	100		